Augusto José Pinto Souto

Neo-endogenous development on rural properties in São Borja-RS

Augusto José Pinto Souto

Neo-endogenous development on rural properties in São Borja-RS

A case study

ScienciaScripts

Imprint

Cover image: www.ingimage.com

This book is a translation from the original published under ISBN 978-3-330-76967-0.

Publisher:
Sciencia Scripts
is a trademark of
Dodo Books Indian Ocean Ltd. and OmniScriptum S.R.L publishing group

120 High Road, East Finchley, London, N2 9ED, United Kingdom
Str. Armeneasca 28/1, office 1, Chisinau MD-2012, Republic of Moldova, Europe
Managing Directors: Ieva Konstantinova, Victoria Ursu
info@omniscriptum.com

Printed at: see last page
ISBN: 978-620-8-62650-1

SUMMARY

DEDICATORY

To God;

To my wife Clàudia and my sons Fernando and Felipe;

To my mother Yolanda (In Memorian);

To my father Fernando.

THANKS

I would like to thank my advisor, Professor Dr. Joao Serafim Tusi da Silveira, who accepted my guidance, even though the field research was already underway, and built this dissertation with his qualified guidance, proceeding in an intelligent manner, using patience and wisdom.

To all the Master's teachers and their staff.

To Professor José Roberto de Oliveira, a partner who made his methodology available and, being helpful, carried out practical actions in the field research for the analysis of the local human and social capacity index in Sao Borja.

To the respondents of the field surveys.

To my colleague Roberto Carlos Dalongaro for his partnership and companionship during the trips.

EPILOGUE

The authors of rural reality:

Respondent No. 235, 25/04/2012 - Wife, mother of children who study and work on the farm. She opted to leave her job as a teacher in order to become a collective entrepreneur on the family farm: "In the municipality, people lack **attitude** and **action.** This is **cultural**. Work is **a habit**!".

Searched N° 257, 19/05/2012 -

Small rural producer, with eight family members on the property. President of an association. The municipality needs a "**local** agricultural policy", with "**support from public authorities**" (federal, state and municipal) and "**universities, teaching and research institutes**".

Searched N° 278, 16/06/2012 -

Rural producer with just over one rural plot. He owns his own land, works in partnership with collective entrepreneurship and his partners are his neighbors. "The **lack of unity** is a **lack of interest**".

Survey No. 287, 18/07/2012 - Dairy farmer, specialized and technified. "Rural areas have a **low cost of living**. But **young people are looking to move to the city**. **Rural education** should include **agricultural practice.**"

We therefore need a new paradigm, a new vision of reality, a fundamental change in our thinking (CAPRA, 1982, p.14).

May this study contribute to a paradigm shift...

SUMMARY

This dissertation proposes an alternative form of neo-endogenous development for rural properties of up to 100 hectares in the municipality of Sao Borja - RS. The region analyzed covers a total area estimated at almost 6,000 hectares, in which 248 properties were sampled and diagnosed. The theoretical and empirical framework used is based on elements inherent to exogenous development, endogenous development and neoendogenous development, especially with regard to the main factors conditioning rural development; and on data and information obtained from surveys carried out on the municipality's 330-year history, on the socio-productive profile of small and medium-sized rural properties and on the human and social capacity for local development. Methodologically, exploratory, qualitative and quantitative research was used, including a wide range of technical procedures: bibliographical and documentary research, *surveys*, case studies, content and citation analysis and descriptive statistics. SPHINX software was used for tabulation and analysis. It was concluded that the critical success factors for local neo-endogenous rural development are a non-individualistic culture, integrated production systems that rescue the skills and traditions of working together, a focus on production/market scale, promotion and incubation (initial push), the establishment of partnerships and the re-establishment of agricultural cooperativism, among others.

Keywords: Rural Development, Neoendogenous, Sao Borja

INTRODUCTION

This Master's thesis in Strategic Management of Organizations is about the neoendogenous development of rural properties of up to 100 hectares in the municipality of Sao Borja - RS.

The region analyzed is the territory of Sao Borja, in which the 248 rural properties are in dispersed locations, surrounded by large private properties and properties owned by large companies in the municipality. The properties investigated include rural men and women, predominantly family farmers: settled farmers, land bank farmers, family farmers in general and, to a lesser extent, rural entrepreneurs. The history of the region under investigation is analyzed, as are the cultural, ethnic, social and political issues that complement each other. A historical survey, research into the socio-productive profiles of small and medium-sized rural properties, and research into the human and social capacity index for local development in Sao Borja were used to gain a broad understanding of the region in question.

The dissertation will consist of the following chapters:

Introduction - Presents the topic, the problem, the hypotheses, the general and specific objectives, ending with the justification.

Capitulo 1 It deals with the theoretical and empirical framework on which the authors are based, providing the foundation for the subsequent analyses.

Capitulo 2 - It records the methodologies used. This chapter aims to externalize how the field research and surveys were carried out, as well as the characteristics of these researches and surveys.

Capitulo 3 - It focuses on the analysis, the results and the proposal. The analysis is made up of tables and plotted relationships of field practices, relating them to theoretical-empirical and historical interactions and resulting in the proposal of this dissertation.

This dissertation concludes with final considerations, reflections on the topic, stimulus for future studies and references. This is followed by appendices and annexes. As for the theme, it refers to rural neo-endogenous development, and the line of research is development and social organization.

The subject is the study of the neoendogenous development of rural properties of up to 100 hectares in the municipality of Sâo Borja - RS.

A municipality with agricultural and livestock characteristics - as well as strong political dominance - has historically moved towards individualism. Individualism presents itself as a threat in terms of social inequalities. And entrepreneurship as an opportunity. In the past, the municipality was strong in cooperativism, but today this is considered a weakness.

The aim is to investigate whether the foundations of rural development meet the requirements to adapt to cultural and ethnic characteristics and social, political and economic factors.

The question is: Would these factors lead to development in the rural segment of the municipality of São Borja studied?

The following hypotheses were raised for this dissertation:

a) The study of the possibilities of rural development for the actors of the community of São Borja is suited to the cultural and ethnic, social, economic and political issues of the territory in question;

b) Cultural individualism is a conditioning factor for local rural development.

As for the methodology used, the research phase was based on an exploratory and descriptive study, using a qualitative and quantitative approach. The technical procedures used included bibliographical research, documentary research and a *survey.* It was a case study and ex post facto research, as well as action research.

The general objective of this dissertation is to analyze the neoendogenous development of rural properties of up to 100 hectares in the municipality of Sao Borja - RS.

The specific objectives are:

- Study development theories;
- Investigate the history of the municipality, its ethnic formation and interrelationships with social, political and economic factors in Rio Grande do Sul, with a focus on Sao Borja - RS;
- Using historical research and field research: research into the socio-productive profiles of small and medium-sized rural properties and research into the influence of human and social capacity on local development in Sao Borja, RS;
- Use the methods of content analysis, citation analysis, lexical analysis in a comparative and integrative way;
- Present results and suggestions;
- Propose a neoendogenous pilot project in the region studied.

This study is justified by the high proportion of rural properties to be investigated, with variables that will have to be analyzed in order to suggest development alternatives. It is also based on generating conditions for research, based on information from surveys of the socio-productive profiles of small and medium-sized rural properties, and analysis of the influence of human and social capacity on local development, for content analysis and the study of development models.

The research environment is a socio-economic and productive segment that requires attention from public policies and the interaction of public, private and community partnerships aimed at the rural development of the municipality.

1. THEORETICAL AND EMPIRICAL FRAMEWORK

1.1 EXOGENOUS AND ENDOGENOUS DEVELOPMENT

According to Dallabrida (2011), exogenous development arose from interventionist approaches that opposed the automatic nature of free market mechanisms and advocated external intervention in the market to combat regional inequalities.

At the same time, neoclassical growth models were constructed which produced characteristics summarized as: the presence of constant returns to scale for labour and capital and diminishing marginal returns for each of the productive resources (DALLABRIDA, 2011).

Endogenous development aims to meet the needs and demands of the local population through the active participation of the community involved. Rather than making gains in terms of the position occupied by the local production system in the international or national division of labor, its aim is to seek the economic, social and cultural well-being of the local community as a whole. As well as influencing productive aspects (agriculture, industry and services), the endogenous development strategy also seeks to act on the social and cultural dimensions that affect society's well-being (BARQUERO, 2001).

For this researcher, endogenous development requires the creation of a favorable institutional and economic environment, provided by the combination of the use of local resources and services, as well as cooperation between the actors, since they are in a position to promote change. Under these conditions, social capital is of great importance for endogenous development.

Dallabrida (2011) apud Vàsquez-Barquero (1998; 2000; 2001; 2002), Albuquerque (2001) and Tenório (2008), associates endogenous development with the capacity of the local community to use the development potential that exists in the territory and to respond to the challenges at a given historical moment. Thus, existing development allows local communities to produce adequately to meet the needs of the population.

Schneider (2006) uses historical roots to characterize the evolution of the endogenous rural development process. For him, this process is the result of the combination of a set of socio-economic and historical-cultural factors which, in the same territory, have led to the development of a productive matrix and a social confrontation which reproduces itself, based on synergies produced by the endogenous process of capital accumulation itself.

In short, it uses the characteristics of the exogenous and endogenous rural development models of Alves (2008):

Table 1- Overview of exogenous and endogenous rural development models

Features	Exogenous development	Endogenous development
Key principle	Economies of scale and concentration	Local arrangements (natural, human &

		cultural). Resources for sustainable development.
Dynamic force	Poles of urban growth. Rural areas are conceived as a source of food and primary products for the expansion of urban economies.	Local companies and initiatives
Function of rural areas	Production of food and primary products for the expansion of the urban economy.	Diversification of economies and services.
Major development problems	Low productivity and marginalization	Limited capacity to areas/social groups from participating in economic activities.
Focus of rural development	Agricultural modernization: stimulating the mobility of capital and labour.	Building capacities (skills, institutions e infrastructure). Overcoming social exclusion.

Source: Alves (2008 p.31).

1.1.1. Main factors conditioning rural development

According to Putnam (2002), social capital depends on voluntary cooperation, like that practiced by revolving credit associations. He also refers to the rules of reciprocity and the systems of civic participation that stimulate cooperation and social trust. The author quotes "Trust is a property of the social system, as much as a social attribute". Stocks of social capital, such as trust, norms and participation systems, tend to be cumulative and mutually reinforcing.

According to Fukuyama (1996) apud Coleman (1988), social capital is the ability of people to work together towards common goals for their groups and organizations. In addition to knowledge and qualifications, a distinct part of human capital has to do with people's ability to associate with each other. He goes on to say that the ability to associate depends on the degree to which communities share norms and values and are willing to subordinate individual interests to those of larger groups. These shared values give rise to trust, and trust, in turn, has great and inestimable economic value.

Still for Fukuyama (1996), social capital, a crucial proof of trust and a critical circumstance for the health of an economy, rests on cultural roots. He also admits that one of the most important lessons of economic life is that the well-being of a nation, as well as its ability to compete, is conditioned by a single, overarching cultural characteristic: the level of trust inherent in society. For this author, considering that community depends on trust, and that trust is culturally determined, it will emerge spontaneously to different degrees in different cultures.

Trust is the expectation that arises within a community of stable, honest and cooperative behavior, based on norms shared by the members of that community (FUKUYAMA, 1996, p.41).

In this definition, Fukuyama relies on both culture and cultural structure, and comes close to the popularly understood meaning of culture, which is an inherited ethical habitus. The vast majority of populations are educated to follow the moral rules of their society as a matter of habit - in family life, with friends, neighbors or at school. Thus, any cultural community, regardless of the specific and implicit ethical rules, will create a certain degree of trust between its members.

When analyzing agricultural activity, Gimenez and Tóffolo (2005) apud Machado et. al. (2000), found that it is generally related to family tradition, where children learn from their parents the tasks, routines, modes of production, lifestyle and standards of conduct in the family and in society in general.

In relation to social capital, Baquero and Cremonese (2008) argue that it can help develop collective identities oriented towards the social and community, providing the materialization of collective goods that are essential for economic development.

Alves (2008) reports that the knowledge produced in rural areas is the product of interactions between traditional knowledge (caboclos, Indians, blacks and European settlers) and fundamentally emerges from scientific research networks that influence public rural development policies, and is disseminated through a network of technical assistance and the marketing of inputs.

In the educational field, Vergara (2010) comments that our culture has favored individual learning. In the family, at school and in the company, individualism has been reinforced. However, even this individualistic learning has not always achieved the desired results. Demonstrating the importance of education in personal, local, regional and endogenous development, they generally refer to the nature of the effects of investments in education by saying that:

> One of the great difficulties in assessing the impact of investments in education comes from the fact that they not only influence the living conditions of those who are educated (private effects of education), but also generate a series of externalities on the well-being of those around them. From a private point of view, education tends to raise wages via productivity increases, increase life expectancy with the efficiency with which existing family resources are used, and reduce family size, with a decline in the number of children and an increase in their quality of life, thus reducing the degree of future poverty. However, we believe that the externalities generated by education can, in general, greatly outweigh its private effects. The magnitude of the external effects of education is, however, little known and difficult to estimate. (OLIVEIRA 2011b p. 46 apud BARROS e MENDONÇA 1997, p. 1)

The IBGE's 2006 agricultural census in Rio Grande do Sul provides the following information in percentages on the number of rural properties and considering one person per unit; these may be owners, tenants, partners or occupants: Adult literacy (1.36%); Incomplete elementary school - 1st grade (71.39%), Complete elementary school - 1st grade (9.66%); High school or complete 2nd grade (6.03%), Higher education (3.22%); No one else can read and write (2.55%) and Cannot read and write (4.23%).

Breitenbach and Brandao (2009) point to the possibility of fostering social capital with "appropriate education" for farmers and the strengthening of local actions. They comment on the criticism of the redefinition of the role of the state by removing responsibility for guaranteeing social rights, and encouraging participation, allowing civil society movements to develop.

Corroborating this is the need for an institutional arrangement to help inter-municipal organizations diagnose the main rural problems in their respective micro-regions, plan development actions and raise the necessary funds to carry them out. In this case, the role of the federal and state governments should be to stimulate initiatives which, in the future, may be self-financing, but which will be difficult and/or slow to emerge if there isn't the indispensable "initial push" (BREITENBACH; BRANDĂO, 2009 apud VEIGA. 2001). This initial push cited by the authors and by Veiga (2001) can be exemplified in initial non-repayable resources, technical assistance with rural extension and after incubation the projects would become self-financing.

Entrepreneurship is important for endogenous development and is a necessary factor for development. Oliveira (2011b) apud Schumpeter (1961, p. 109-110) offers alternatives for entrepreneurship:

> New opportunities are constantly being offered by the environment; in particular, new discoveries are continually being added to the existing stores of knowledge. Why shouldn't the individual make use of both new and old possibilities and, according to his understanding of market conditions, raise pigs instead of cattle, or even prefer a new crop rotation if this seems more advantageous to him? And what kind of specific new problems or phenomena, not found in the circulatory flow, might arise there?

In the opinion of Antunes and Ries (2001), individual rural entrepreneurship is only successful when it is geared towards the future, innovation and creativity. When there is a willingness to change and to pay attention to needs and opportunities. When you think of new alternatives, both in the technical sense (new crops, management, inputs, etc.) and in the administrative sense (marketing, human resources, investments, machinery, etc.). And when they are ready to take advantage of the alternatives that arise. Rural entrepreneurs have the attitude to take action.

Regarding rural collective entrepreneurship or extra-entrepreneurship, Julien (2010) apud Bygrave (1989), Aldrich (1990) agree that their definitions are insufficient when classifying individual entrepreneurship, when in fact the classification interrelates the general environment and relationships, such as family, networks, overcoming the definitions of individual role models.

When commenting on the success of a rural enterprise, Zuin and Queiroz (2006) apud Cella and Peres (2002), extrapolate the simple notion of maximizing profit in activities, linking subjective elements, such as the need to fulfill the entrepreneur's desires and aspirations. They also include the involvement of elements such as family and social life, decision-making autonomy and community recognition. These authors identify the following characteristics inherent in successful rural producers:

- Sharing the risk of agricultural activities;
- Use of operational and long-term planning;
- Use of expense control, both for the activity and for the family itself;

- Incorporation of subjective elements, such as freedom and decision-making autonomy;
- Information and communication skills;
- Involvement of family members, both inside and outside the property;
- Concern for community affairs, social recognition and the preservation of the environment;
- Cultivation of the personal and family dimension that directly interferes in the management activities of the rural enterprise.
- Strong cultural heritage, with an emphasis on work ethic and a greater stock of social capital.

Julien (2010) explains regional entrepreneurship, and proposes a pyramid comprising four triangles in the logic presented in the book. The first triangle represents the three elements that form the basis of endogenous entrepreneurship: the entrepreneurs or catalysts of entrepreneurial activity; the organization, as a complement and supplement to their actions; and the environment, which explains the dynamism and benefits in return. The second triangle, the one in front, links entrepreneurs to the environment, and therefore to the economy, where they find a market and resources, depending on the sector in which they operate. The third triangle, on the left, links entrepreneurs, the environment and time. It shows that time acts on the behavior of entrepreneurs, who make choices according to the opportunity and the period. Time also explains the particular evolution of the environment and its dynamism. Finally, the last triangle, at the bottom, takes up the links between entrepreneurs, organization and time, to show that the first two are transformed considerably by time, succumbing to it or taking advantage of it; it also points out that time gradually leads entrepreneurs to become managers because of their risk-taking attitude (figure 1).

Figura 1 - The pyramid of entrepreneurship

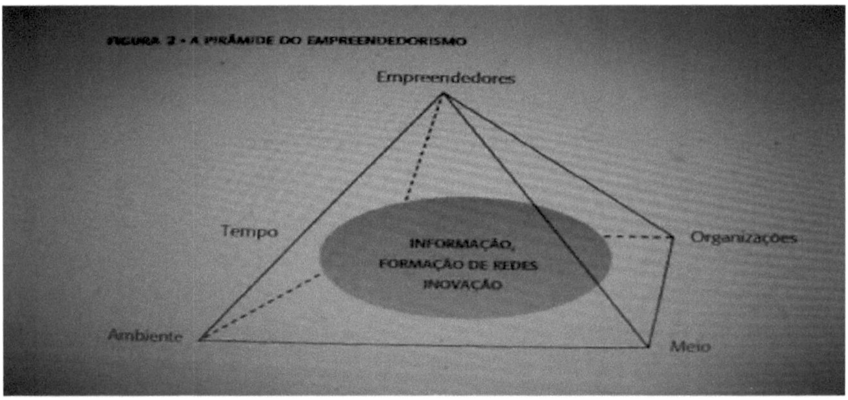

Source: Julien (2010, p.27), photo by the author.

For his part, Oliveira (2011b) apud Schumpeter (1961) presents the specific problem of leadership, which arises with the figure of the leader. The role of the leader is to "do something", applying ephemeral or lasting forms of leadership, which condition endogenous development.

Leadership is defined by Wagner III and Hollenbeck (2002) as the use of symbolic and non-coercive influence to direct and coordinate the activities of the members of an organized group, with a view to achieving its objectives, acting in an autocratic, democratic or liberal manner.

Regarding "Power", March (2010) conjectures that different people want different things or different identities; and that individuals and groups compete and cooperate with each other, trying to satisfy their individual preferences and identities. For him, power is the ability to get what you want or to achieve a certain identity.

In Morgan's view (1996, p. 163) "Power is the means by which conflicts of interest are ultimately resolved. Power influences who gets what, when and how".

Khanna's (2009) comments on politics, politicians and power are also interesting. For him, what counts is what that person does for you, your street and your neighborhood. Politics is very personal and local, often shaped by feudal behavior. Thus, the politician becomes an all-powerful person who can solve things, further strengthening the paternalistic/clientelist system.

According to Morgan (2006) apud Weber (1947), domination can occur in various ways. Charismatic domination occurs when a leader dominates by virtue of his or her personal qualities. Traditional domination occurs when followers accept the leader's command as a custom or acquired right, as in inherited leadership. Rational-legal domination occurs when laws, rules, regulations and procedures legitimize the power of command, as in a bureaucracy.

Rosa (2011) states that the state of Rio Grande do Sul is an economy in transition in the sense that a movement is underway in which activities that are intensive in created skills are gaining a share in the productive structure and activities that are intensive in natural skills are losing out. The latter remain the most important for output, exports and employment. Natural capacities use land, irrigation water, electricity, infrastructure, machinery and equipment, and require roads and storage. Porto (2007) cites studies by the United Nations Agricultural Organization which emphasize three factors that determine physical productivity per hectare: a) genetics; b) production technology with equipment, characterized as the "green revolution"; and c) irrigation, as a way of avoiding losses and boosting productivity, using water as a factor in production efficiency.

Other aspects of development should be highlighted. Oliveira (2011b) apud Boisier (2005, p.32) emphasizes that every organization, whether business or territorial, must think in the long term. It must answer four questions: "1) What to produce? Where to sell? 2) What projects to develop? How to finance them? 3) What human resources do you have? What can they be employed in? and 4) What corporate image should be built? How to market it?".

Family farming that is highly integrated into the market, capable of incorporating the main technical advances and responding to government policies cannot be characterized as peasant (ABRAMOVAY 1998, p.22).

Regarding rural activity on small properties, the author states:

> Agricultural activity has been a source of dynamism and wealth for some regions and stagnation for others. In this respect, there is no way for small farms to prosper by producing cattle and/or beef, however productive they may be. As a general rule, there is a need for a productive reconversion of the grain-producing smallholding segment towards activities such as fruit-growing, horticulture, floriculture, dairy farming, small animals and genetics (ROSA 2011, p.42).

Rosa (2011) states that from a technological-productive point of view, there are no major difficulties in converting the production of small farms that currently produce grain towards high value density activities such as fruit growing, for example. There are still major economic difficulties related to the producer's insertion into the market. And this market insertion is either directly with the final marketing link for products consumed in natura or with the industrial link for processed products.

In both cases, according to Rosa (2011), the production potential of regions with a predominance of smallholdings far exceeds the absorption capacity of the local markets in which they operate, and in most cases there is no local industrial capacity. He asks, why is there no industrial capacity installed in most regions? The answer is simply because they have production potential, but no supply to justify the installation of industrial plants which are, as a rule, demanding in scale.

He concludes that the regions don't produce fruit, for example, which generates many times more value and profit per hectare than grain and employs many times more labor units, because they don't have a market. But the regions don't have a market because they don't produce the fruit in the volumes that would make market creation viable. This means that the majority of regions where smallholdings predominate are trapped in a vicious circle that condemns a significant proportion of their inhabitants to poverty (ROSA 2011).

He points out that in livestock farming too, producers need to work to break out of this vicious circularity. He emphasizes that this is not a trivial undertaking, but it is possible to move forward with success stories in Brazil, the world and RS; there are many and financial and material resources were not always the most important for the solutions and results achieved. Most of the time, a new way of thinking about regional issues and a cooperative effort were the decisive instruments. In RS, this is no different. These two instruments, if properly mobilized, can transform the vicious circle of poverty into a virtuous circle of prosperity and well-being for the producers involved and their communities (ROSA 2011).

Sen (2012) explains that the best way to end misery is to end the cause of misery - the real deprivations that affect human lives.

Figure 2 shows Porto's (2007) vicious circle of poverty. This figure shows the relationships and negative constraints on rural development.

Figura 2 - The vicious circle of poverty

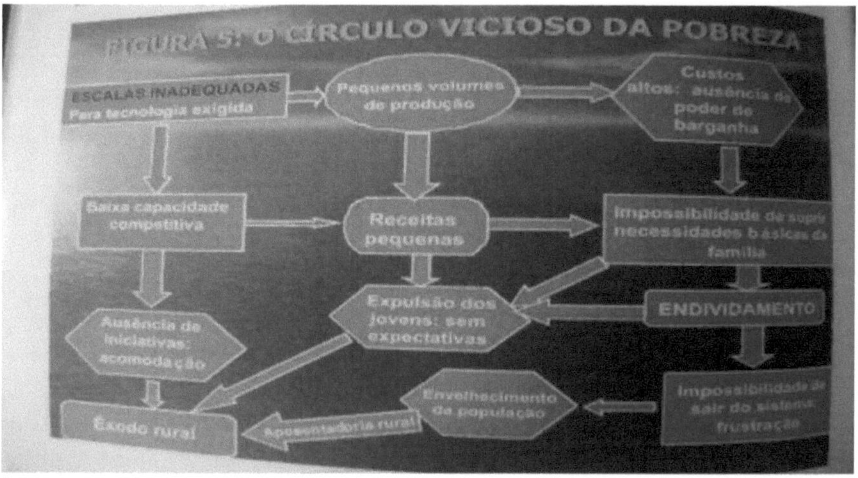

Source: Porto (2007 p.118), photo by the author.

According to the Institute for Applied Economic Research (2003), the population living in poverty does not earn up to two national minimum wages, which is considered a minimum to survive with dignity.

In the study by Waquil; Mattos (2002) in the article "Public policies, family farming and rural poverty in Rio Grande do Sul", they conclude that the general income situation in rural areas is worse. They point out that a third (between 32.89% and 37.76%) of the poor are located in rural areas. And the poverty line, two minimum wages in 2012 (R$ 622.00 x 2 = R$ 1,244.00) corresponds to 18.34% of the population. The authors point out that in addition to income, housing conditions, schooling and health are also important, with a focus on comparing living conditions and poverty situations between rural and urban areas.

In fact, the biggest difficulties for small rural producers are concentrated in the "before and after the gate", i.e. small rural producers pay more for inputs because they buy small quantities and sell their products for lower prices due to the small volume, in addition to not having the means of transportation, thus being in the hands of middlemen (SEGATTI, HASPANHOL 2008, p.12).

We therefore need a new paradigm, a new vision of reality, a fundamental change in our thinking (CAPRA, 1982, p.14).

Zonta (2012) reports on ATER Law No. 12.188/2010, which establishes the National Policy for Technical Assistance and Rural Extension (PNATER) and defines the objectives of ATER services. The National ATER Seminar (MDA/DESER 2008) - Technical Assistance and Rural Extension, which was the basis for the law, produced the final report which analyzes: the existing difficulties and proposals with a focus on Rio de Janeiro.

Grande do Sul, in Table 2:

Table 2 - Difficulties in the ATER service and proposals for RS

Difficulties in the ATER service in RS:	Proposals for RS:
• Deficient and insufficient ATER; • Substitution of inputs; valorization/commercialization; lack of research. Change of attitude (technical/teaching); experiences lack integral public articulation; • Commercial production has led to indebtedness; difficulties in understanding and serving different audiences (discontinuity of state policies); • Monoculture leads to impoverishment; bureaucracy in agreements (delays); poorly equipped institutions; • Lack of resources; too many families for one technician to attend to; while non-governmental organizations have difficulties accessing resources and carrying out projects; • The invisibility of women; work is segmented in the family; • Poor functioning of the Sustainable Rural Development Councils in the municipalities and of the rural movements; councils that are not very representative or functional; • Organizational level of entities still precarious (dependence on the state); family farming has little effective political representation;	• Ensure that there is a political decision to face up to the challenges of implementing food sovereignty and security in the country. This initiative must be adopted by all development agents working in the family farming sector; • Increase the number of professionals; there is a lack of extension agents; • Qualifying Pronaf; priorities of investing in those who need it; • Encourage social movements to increase their support and coordinate ATER actions; Establish priorities/invest in those who need it; • Harmonize the institutions; increase resources (financial, technical, material); reduce bureaucracy in the release and application of resources; • Qualify and prepare ATER for working with young people and women (the family as a whole); valuing and preparing young people and parents; • Training councilors, managers and local leaders (meetings, seminars, conferences, etc.); - Intensify the work of
restructuring the councils; - Destructuring of universities due to lack of students; school curricula do not include rural areas, emphasis on education in the countryside.	children/schoolchildren; valuing family farming (when young people have to do an internship, it is not considered family property); working more with families in the communities; intensifying support for the pedagogy of alternation; - Consider the producer as an actor in the process; forums to discuss Brazilian rurality (in family farming); intensify work on territories; work on the issue with young people and schoolchildren; prepare ATER agents (value participatory methodologies).

Source: MDA/DESER (2008), adapted by the author.

The manager of the IBGE's geography coordination, Figueiredo (2012), says: "There are major difficulties because of the lack of training for producers. Even in the simplest agriculture, the market will increasingly demand qualification and technological innovation". In the same article, , the journalist interviews the Superintendent of the National Rural Apprenticeship Service in Rio Grande do Sul (Senar-RS), who comments:

> These are people whose work demanded so much of them during their youth, and who didn't have the opportunity to continue in formal education. We're trying to fill a gap with training that farmers can use in their rural activities. Education is still an obstacle in rural areas. The training of producers has not kept pace with the evolution of technology (TIETBOHL, 2012, p.16).

Pereira (2006) explains what empowerment is. Empowerment, in general, means the collective action taken by individuals when they take part in decision-making spaces, and social awareness of social rights. This awareness goes beyond taking individual initiative to understand and overcome a situation (reality), to understanding complex networks of social relationships that inform broader economic and political contexts. Empowerment also enables individual emancipation from the collective consciousness needed to overcome "social dependence and political domination". Not just an instrumental concept of efficiency, the idea of empowerment must play an important role in the possibilities of realizing people's rights.

According to Wilkinson (2008) logistics, in particular transportation and planning, affect agricultural productivity. The author continues that by adopting a strategy, in the smaller modules, of cooperativism, the cooperative defines different priorities for technological and organizational innovation. It produces specific patterns of coordination and cooperation between producers with efficiency and competitiveness, not at the individual level but in the global organization of production and distribution logistics.

Valle (2010) comments on the need for integration and partnership between rural producers, government organizations, teaching and research institutions, technical assistance and rural extension to meet market demands. Through the Good Agricultural Practices Program (GAP), the main objectives of this program are to increase profitability and improve the competitiveness of production systems, as well as to ensure that the consumer market is supplied with safe food. The organizations collaborate in the diagnosis of farms, with training and capacity building for rural producers.

An alternative is presented:

> Certifications of regional origin or geographical indications are awarded to products from a certain region, produced under specific conditions, and which have recognized quality standards. Products from these origins must be renowned and comply with production standards, and basically have characteristics defined by the influence of nature and, to a lesser extent, by the influence of man. Geographical indications provide good remuneration for producers and are usually a key differentiator in their quality of life and professional success HERTER (2011, p.78-79).

Castells (1999) analyses that in the mode of production, capitalism, which aims for profit and statism, aims or aimed to maximize power. From the 20th century onwards, development would be characterized in three modes: the agrarian mode, the industrial mode and the informational mode of development. Informationalism is based on knowledge and information technology, in which there is an intimate link between culture and productive forces.

1.1.2 **Neo-endogenous rural development**

This section presents the theme of this dissertation. Neo-endogenous development: endogenous development and exogenous development complementing each other, the local-global.

Recently, Alves (2008) apud Buttel (1994), demonstrates the need to integrate endogenous and exogenous models, because according to him neither of them, in isolation, is capable of explaining rural development processes. The exogenous perspective is incapable of explaining micro-social processes. And the endogenous approach is incapable of explaining macro-social processes. Under these conditions, combining the two would provide a theoretical perspective capable of grasping the complexity of socio-rural reality.

The author also proposes interweaving the models of globalization and re-localization. His argument is that micro-social theories, which emphasize the specificities of endogenous development, have shown that local factors do not disappear within the dynamics of globalization, but are transformed. The fundamental argument is that the globalizing proposal overvalues homogenizing forces (technological modernization and markets), neglecting the diversity of techniques and arrangements that exist in the local space. In addition, re-localization highlights various issues, such as the fundamental role that local knowledge, the intrinsic complexity of coordinating agricultural tasks and chores, gender issues and domestic aspects of the organization of agriculture - play in the dynamics of agricultural and rural systems and societies in general.

In view of this, Alves (2008, p.34) apud Ward et. al. (2005), suggest the nomenclature neoendogenous to replace the concept of endogenous development. For these authors:

> [...] the neoendogenous approach is inspired by theories which maintain that the key to local development is the construction of a local institutional capacity which is both capable of mobilizing internal resources and dealing with the external forces at work in a given region. This perspective emphasizes that not only economic or business development needs to be incorporated into the region, but that the way to achieve this is through the participation of local actors in internal and external development processes. They also propose studying how the local level interacts with the extra-local level. In the view of these researchers, the critical point is how to increase the capacity of the local level to use wider areas, resources, actions and processes to its advantage.

Becerra and Bravo (2009) present a new methodological plan, which they conceptualize as neo-endogenous rural development. This model is based on three pillars. The first is based on endogenous development. The second is the development of social capital (endogenous units, local/global actors and neoendogenous units) and the third is the promotion of participatory democracy. The authors' methodology proposes a multisectoral and transversal approach. They aim to develop endogenous potential through innovation, the participation of actors and the validation of their social capital.

1.2 SÂO BORJA AND THE SECOND FOUNDATION OF SETTLEMENTS - SEVEN PEOPLES OF

THE MISSIONS

Padilha; Trentin (?) comment that in the first phase of the Jesuit-Guarani civilization, between 1626 and 1637, the Jesuits, in the Tape missions, were interested in teaching the fundamentals of the Christian religion to the Indians, and with the bandeirante invasions; the Jesuits and the Guarani were forced to migrate to the right bank of the Uruguay River and stayed 45 years away from the territory of Rio Grande do Sul.

Langer (2005) informs us that in detriment to the advances of the Portuguese (bandeirantes), the Spanish colonial authorities reacted. It was as a defensive measure that the Society of Jesus resumed the founding of missionary settlements on the left bank of the River Uruguay, which had been abandoned since 1640, giving rise to the so-called Seven Peoples of the Eastern Band between 1682 and 1707.

Bruxel (1978, p.30) mentions the Seven Peoples: Saint Nicholas -1626, Saint Michael (Archangel)-1632, Saint (Francis of) Borja- 1682, Saint Louis (Gonzaga)-1687, Saint Lawrence (Martyr)-1690, Saint John (Baptist)-1697 and Saint Angelo (Custodian)-1707.

In 1682, the Jesuits and the Guarani crossed the Uruguay River again to begin what is known as the Second Phase of the Missions in Rio Grande do Sul. The first took place between 1636, when Father Roque González de Santa Cruz founded Sao Nicolau do Piratini, considered the First Querencia of Rio Grande, and 1638, the year in which the Jesuits and Guarani abandoned their 18 Povos that they had created on the left bank of the River Uruguay, due to the attack by the São Paulo bandeirantes.

The return of the missionaries to present-day Rio Grande do Sul began in Sao Borja, which is why this city is known as the First of the Seven Peoples. The founding of Sao Borja had the military objective of fighting the Portuguese who had recently founded the Colony of Santissimo Sacramento (VENTURINI, 2009, 115).

Padilha; Trentin (?) explains that the reducion of Sao Borja originated from a ranch organized by the Guarani of San Tomé (Argentina, right bank of the Uruguay River). The founder was the Jesuit Francisco Garcia de Prada, due to the large number of Indians who migrated from San Tomé, setting up the reducion of Sao Francisco de Borja. This was made up of the "charrua" ethnic group and according to Freitas (1935) they were Guenoas, Charruas and later Guarani.

Flôres (2012) formulates a hypothesis in his book about the emphasis on the expulsion of the Jesuits who lived in the indigenous reductions, one of which was Sao Borja, which resulted in a period of economic, political, socio-cultural and demographic disarticulation, a process that occurred predominantly between the years 1756 and 1801.

Figure 3: The thirty peoples

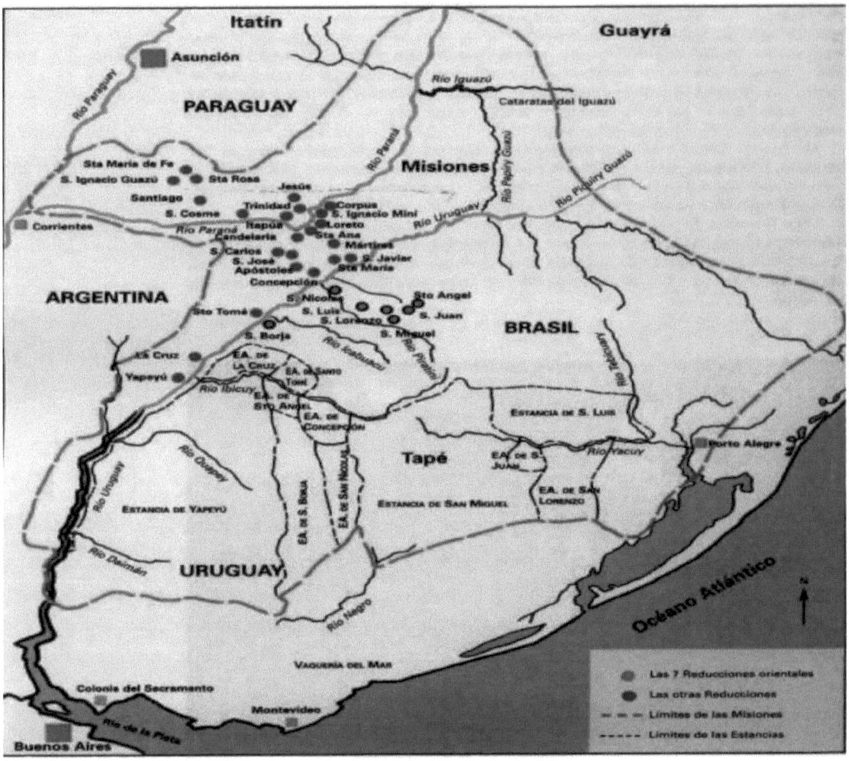

Source: PADILHA, TRENTIN [?], p.9 apud ABOU, 1996, p.110.

1.2.1 **The people of Sao Borja**

The purpose of this section is to explain the history of Sao Borja from its founding to the 21st century and its complementary features.

1.2.1.1 From the formation of the Sao Borja people to the 17th century

Sempé (1982) comments that the founder of Sao Borja in 1682 was Father Francisco Garcia de Prada, born in Spain.

In the Sao Borja tricentenary collection, Sempé (1982, p.94) reports a different fact:

> But in 1682, as we have seen, there was only the DIVISION. The transmigration that led to the establishment only took place in 1687. And Father Francisco Garcia's letter, dated December 1683, makes no reference to Sao Borja - whose DIVISION had taken place the previous year - and tells us quite clearly about the impossibility of establishing a Redución among the Guenoas, a subject dealt with in extensive research that we presented on the occasion of the III Symposium .[1]

1 Moarci Matheus Sempé - "O Padre Francisco Garcia e a fundaçao de S. Francisco de Borja" - Proceedings of the III National Symposium of Missionary Studies - Dom Bosco Faculty of Philosophy, Sciences and Letters

Flôres (2012) comments that in the work of organizing the Sao Borja Reduction, the charruas were of great value in helping with pastoral activities due to their extensive knowledge of the areas of the cowsheds, located on the plateau, from where they removed the cattle for the subsistence of the reduced populations.

In the magazine of the Instituto Histórico e Geographico e Ethnographico do Brasil (1863, p.711 and 712), he explains about São Francisco de Borja:

> Sao Borja, which is the southernmost of the Jesuit settlements on the eastern bank of the Uruguay River, was built twenty leagues south of Sao Nicolào, and a league and a half north of this river on the western side was the settlement of Sao Tomé, which in 1699 had had a colony founded there with Charrua Indians, the other six eastern settlements being made up of Guarani. During the time of the Jesuits and the Spanish, the town of Sao Borja was the most commercial post on both banks of the Uruguay River. [...] Shortly before the change of dominion, Sao Francisco de Borja suffered an invasion by Minuans, which caused serious damage and was partly responsible for its delay.

Rodrigues (1982) reports that the economy of the reduction was based on the agro-pastoral industry and yerba mate. The breeding fields belonging to the people of Sao Borja were separated; the headquarters fields, between the Icamaqua and Butui rivers up to the Iguaraçà mountain range, were divided into two estancias: that of the reducao and that of Sao Gabriel at the bottom, plus one more establishment - the chapel of Sant'ana.

Bruxel (1978, p.117 and 118) writes about the estancias in his book:

> Cattle, horses and wool were raised separately, to make work easier and avoid disturbances (wild horses would be a danger to pregnant sheep and cows). [...] The annual gathering of the cattle, which could take up to two months, was never carried out by private Indians, but always by the community, which entrusted some 40 to 50 muleteers with the task.
>
> Livestock was of the utmost importance to the Mission Indians. Even in times of agricultural abundance, beef was the main food; in times of scarce harvests, it was up to half the breed or more. It was therefore difficult, if not impossible, to sustain the Reductions without an abundance of beef.

Bruxel (1978) comments that an expedition of volunteers went to the herb fields every year to harvest enough grass for the daily consumption of all the families in the Reductions and also some for export. Sao Borja had a herb plantation near Lagoa dos Patos, in a place called Serra do Erval.

1.2.1.2 People of Sao Borja in the 18th century

Langer (2005) talks about the definition of the Iberian colonial boundaries, established by the Treaty of Limits in 1750, in which Portugal handed over the Colony of Sacramento to Spain in exchange for the Seven Missions on the left bank of the River Uruguay. The author continues that for the Guarani of the Seven Peoples, Article XVI represented a supreme and inconceivable punishment. This agreement established nothing less than the surrender of the lands and settlements built up over

- Santa Rosa, RS, Oct.1979 - with separata.

decades to their ultimate enemies:

The people of Sao Francisco de Borja were mobilized to carry out the transmigration, so they were not included in the governor's threats; on the contrary, they were pointed out as an example to be followed by the others (LANGER, 2005, p. 83).

Maurer (2010) comments that of the eastern peoples of the Uruguay River, the only one not to take part in the battle was San Borja. This issue has been observed in several works, in this vein Rabuske (1985, p.129) states: We can certainly say that fortunately the "Borjistas" were not present or did not take part in Caibaté on February 10, 1756. They did, however, arrive after the battle.

In June 1753, the Spanish commissioners Andonaegui and Altamirano sent important notices to the Missions. The purpose of these notices was to press for changes that had to be made by August 15th of that year, otherwise war would be declared on the Seven Peoples. The letters were addressed to all the Jesuits and the six curas of the six peoples, because they had already obtained the obedience of the Sao Borja Mission (MAURER, 2010 apud GOLIN, p. 268, 2004).

In October 1756, Pedro Cevallos was appointed to replace Governor Andonaegui, who had been denounced as inept by the commissioner Marqués de Valdelirios. He went to the missions and discovered that the Jesuits had not incited the Indians to revolt. In the town of Sao Borja, he set up a court in the square in front of the church and, together with Valdelirios, Viana and the other military authorities, began to interrogate the chiefs and mayors of the missions through two interpreters. The Indians were unanimous in confirming that they knew the King's rules and that they didn't move because they would starve to death elsewhere, as they wouldn't have time to plant crops or drive cattle. They also replied that the priests had always begged them to obey orders. Finally, they recognized defeat as a punishment for their disobedience (MAURER, 2010 apud FLORES, 1986).

According to Flôres (2012), the sequence of the Castilian history of Sâo Borja is marked by the dismantling of the reducción system, motivated at first by the Guaraní Wars, which involved the Iberian demarcation forces, the indigenous people and, from 1767, the expulsion of the Jesuits. The period was characterized by the Spanish military and administrative presence, which resulted in the beginning of private occupation of the mission area.

Langer (2005) explains that on the occasion of the administrative changes implemented in the reducciones with the expulsion of the Jesuits in 1768, the Guarani made the following statement: we are not slaves and we want to make it clear that we don't like the Spanish custom which wants everyone to look after themselves instead of helping each other in their daily work[2] . This statement indicates that the Guarani missionaries were aware of, but reluctant to accept, the directives of colonial mercantilism. If the Jesuits' efforts to prevent the introduction of a market economy in the reducciones resulted in failure in this regard, it was because the Guarani were repulsed by any form

2 The author cites the work of Meliâ as fundamental. MELIÂ, Bartolomeu. Potyró: La Cuestión del Trabajo Indigena Guarani. In. Proceedings of the VII National Symposium on Missionary Studies. Santa Rosa: Don Bosco Faculty of Philosophy, Sciences and Letters, p. 295-326, 1990.

of work that lacked a collective character. The author states that there is practically unanimity among scholars that the economic basis of the reducción model, characterized by significant agricultural and extractive productivity, lay in the communal organization of work and property. Private ownership of land and individual work were non-existent categories in the socio-economic mentality of the pre-colonial Guarani. The private family property - abambaé - which the Jesuits endeavored to establish in the mission villages was neglected in favor of communal property - tupambaé. The predominance of communal work and farming over the European family model is a consensus in studies on this subject.

Rillo (1982) asks who administered Sao Borja after the expulsion of the Jesuits? These Spanish administrators, in Sao Borja and in the other Pueblos, treated the Indians like slaves, took possession of the material goods that still remained, did not impose correct methods of administration, to the point that countless Indians returned to their primitive lives or began to work on the first ranches that were planted in the semi-wild interior.

After the Jesuits left the province of Paraguay, the Spanish government concentrated the entire province of the Missions under one jurisdiction. The Indians continued to live in communities. The Guarani work one week for the community and another week for themselves. After the expulsion of the Jesuits in Buenos Aires, the post of Intendant General of the Misiones was created. This was to centralize the collection of profits. They received products to trade and deducted the tithe. The royal treasury received little, despite orders and complaints from the Madrid cabinet. Governed harshly and sometimes forced to work in a way that was harmful to their health, they quickly became disgusted with this new regime and gradually fled their villages. The administrators sometimes abused their authority by inflicting heavy punishments, poor quality and insufficient rations. The agents of the Spanish government and their jobs were for enrichment, not for the general good (RIHGEB, 1863).

1.2.1.3 Sao Borja and the 19th century

In 1801, the Badajoz Treaty was signed, ending the war between Portugal and the Franco-Spanish coalition. On incorporating the missions into Brazil, the people of Rio Grande found a semi-nomadic, culturally helpless population in those fields. In the absence of stimulating cultural interactions, the Guarani were in cultural agony, without the strength to resist the mestizo, whether he was from Corrente, Uruguay or Rio Grande do Sul (CESAR, 2002).

Freitas (1935) comments that José Borges do Canto, a deserter from the Dragoes regiment, and Manoel dos Santos Pedroso, a corporal, and 47 men were given the mission of helping the people of Sao Lourenço, who, eager to free themselves from the Spanish yoke, also took Sao Joao. Sao Miguel on August 13, 1801. Pedroso was on his way to take the Uruguay Passes, in order to prevent resources from coming from the Governor General of the Missions, when he "met a deputation of the main inhabitants of the People of Sao Borja who, leading and manacling their administrator, had declared themselves in favor of Portugal. Pedroso, after having fortified the Passes of Santa Maria,

Santo Izidro, Sao Lucas and Sao Marcos, on the River Uruguay, entered without opposition from the people of Sao Borja".

Flôres (2012) explains that in addition to the military force along the border, sesmarias were distributed among the main lideres who commanded the occupation of the Missoes. The aim was to ensure the sustainability of the settlement over a large geographical area. The economic survival of the Portuguese who settled there was based on the herds that still existed, the production of jerky for consumption and trade, as well as the use of some land to plant vegetable gardens, fruit trees, corn, wheat and beans. Cottage industry appeared with culinary products and the manufacture of clothes.

The mission territory was annexed to the municipality of Porto Alegre, and remained so until 1809, when it became part of the Rio Pardo term by the provision of October 7, 1809 (RODRIGUES, 1982).

> Through the sesmarias, the Portuguese conquered and populated lands that belonged to the Spanish crown. The Portuguese advance was at the expense of the missionary estancias. The sesmaria was an area about one league wide by three long, granted by the King of Portugal to anyone who proved that the land was vacant and owned goods such as slaves and cattle. (...) The authorities granted sesmarias in preference to military personnel. (...) Unfortunately, the distribution of land in the province was vicious. Many poor families wandered from place to place without finding jobs. That's the mystified gaudério (FLORES, 2004, p.12 and 13).

The relentless struggle in the missions. Cesar (2002) reports that in 1811, a new series of fights took place with D. José de Artigas. The caudillo's troops used their low instincts against the population of Sao Borja. In reprisal, Chagas Santos severely punished his opponents, the troops of Andresito Artigas, Artigas' adopted son and born in Sao Borja of Guarani origin; both in the Sao Borja pass and across the river in Sao Tomé. He reduced Guarani villages to rubble, drove away stable and nomadic populations and decimated herds.

The naturalist refers to the houses around the square, which are no longer occupied by Indians, but by whites who pay rent for them, some of which are used as stores. For a hundred white families, there is only one Indian family in Sao Borja, which is now considered a war square. The small number of Indians who belong to Sao Borja live in miserable, sparse huts in the village. Other huts, inhabited by the wives of the soldiers, are equally destitute. Most of these terrible dwellings are built of straw. A hammock, a few jiraos, a copper kettle and a few pots made up the furnishings, and in only two or three of them had a few cornstalks been planted (SAINT HILAIRE, 2002).

Flores (2004) comments on the recently conquered Missôes region, which was divided into large estancias (ranches) with Indians as peasants and muleteers. The trails of carts and mule troops cut across the campaign in different directions. The municipality of Sao Borja, which encompassed the Missôes region, was home to the oldest urban center in Rio Grande do Sul, dating back to 1682 and always inhabited. He points out that the Missôes region produced cotton, corn, tobacco, sugar cane, yerba mate, pears, figs, plums, quince, melons and vegetables. Their estancias raised mules, horses and cattle.

Rillo (1982) reports on the second invasion suffered by Sao Borja. He reports that under the inspiration of the caudillo and brigadier Frutuoso Rivera, who had invaded Rio Grande territory some time before. An important column, commanded by Bernabé Rivera, marched against Sao Borja, then the headquarters of the General Command of the Missions, laying siege to it in the last days of April 1828. Sao Borja was abandoned to its fate by the commander of its forces and its general administrator. After plundering all the people of the Missôes, including Sao Borja, their houses, estancias and churches. Rivera was carrying no less than a troop of 20,000 cows, as well as a war booty of incalculable value, loaded onto 60 ox-drawn carts.

The town of Sao Borja, being the seat of the general command, with the presence of the garrison forces and its proximity to the Uruguay and trade along that river, grew in such a way as to exclude the indigenous population from the urban limits, being replaced by the white population. Since then, it has been impossible to increase the number of other reductions (SILVEIRA, 1979, p.104).

The National Guard was created on August 18, 1831, by the Liberals, according to Flôres (2012) (apud Leal, p.20-21). This was a concern of the House of Representatives to promote the organization of the Municipal Guards.

Rodrigues (1982) reports that Sâo Borja continued under the system of Military Commands, by resolution of the Governor of the Province, Dr. Manuel Antônio Galvâo, dated March 11, 1833, Sâo Borja was dismembered from the municipality of Rio Pardo and until its installation as a Municipal Term on May 21, 1834. It became the town of São Francisco de Borja.

Dreys (1990), during the Farroupilha Revolution. He comments that S. Borja, although generally inhabited by Guarani Indians, among whom there are some whites belonging to the civil and military authorities or merchants. With comparatively elegant architecture, that is, in reference to the general state of the art in the country; a college that was one of the former priests of the company, and a hippodrome.

In the Farroupilha Revolution of 1835 to 1845, the Sâo Borja City Council supported the movement led by Bento Gonçalves da Silva, only to repudiate it and support the imperial and legalist forces (RILLO, 1982).

Belgian traveler A. Baguet (1997), who visited the town of São Borja, reports that the commander housed the team in the old Jesuit school, planted with magnificent orange trees. Almost all of those beautiful trees had been felled to make firewood.

Aimé Bonpland, a wise botanist like many others, once a traveling companion of the illustrious Humboldt, lived miserably in that village, practicing medicine and operating an apothecary's shop. Bonpland, like so many others, fell victim to the cruelty of the dictator Francia, whom he kept imprisoned for many years (BAGUET, 1997).

Flôres (2012) highlights the example of the magistrate, Judge Francisco de Sá Brito, who was considered to have "exceptional intelligence" and "sound culture". One of the first steps he took was to obtain a list of citizens suitable to hold judicial office, which, for the time 1820-1830, was not very easy to achieve because most "good men" did not have an education.

Silveira (1979) notes that at the repeated request of the government, the Legislative Assembly of the Empire decreed, in article 36 of the budget law of October 21, 1843, that the estancias of the Indians of the Eastern Missions be incorporated into the national patrimony. This law had just stripped the poor Indians of the few possessions with which they could still return to a more regular life when the civil war in Rio Grande do Sul ended. This law had already found the few lands of the Indians' former patrimony in the hands of third parties and, in more than half a century, they have not been bothered with a legal claim.

Livestock farming, the economic base of the municipality, developed greatly. In 1858, before the dismemberment of Itaqui, we had 250,000 cattle, 180,000 horses, 30,000 sheep and 1,000 merino mongrels, and many mule pens, which grazed in the fields of Sao Borja and Itaqui and were taken to the Sao Paulo market (RODRIGUES, 1982).

Silveira (1979, p.109) reports: In 1856, the first public teacher, Felisberto Batista da Costa, arrived in Sao Borja, with more than a hundred pupils in his class. Two years before that, a girls' primary school had been set up there.

In 1860, steam navigation began on the river line between the ports of Salto, Concórdia, Uruguaiana, Libres, Itaqui, Sao Borja and Santo Tomé (RODRIGUES, 1982).

Flôres (2012) in his research into the contents of the Minutes of the Chamber of Councillors found indications of the presence in Sao Borja of economic production aimed at food subsistence. The data presented in the Book of Offices (1861-1866) shows that there were 20 mills in the town in 1861. This suggests that there was local wheat production, or that it was the result of the smuggling of Argentine wheat production.

Freitas (1935) reports the invasion by Paraguayan forces, the third invasion of Sao Borja. This took place on June 10, 1865 at 10 o'clock in the morning, when troops and carts were spotted covering the entire stretch between Santo Tomé and Formigueiro. A total of 2,670 Paraguayans and four gunboats landed during this first action. It was necessary for the families in the town to have time to leave.

Figure 4: Fighting in Sao Borja against the Paraguayans

Source: Boeira and Golin (2006, p.111).

Gay (1980, p.153) cites the passage of Dom Pedro II through Sao Borja:

His Majesty walked until nightfall through the Pass of Sao Borja, expressing satisfaction at seeing large plantations of laranjeiras, and questioning various people about the unfortunate event of the enemy's passage on June 10, 1865. On September 28, 1865, His Majesty disembarked at 6 a.m. in the town of Sao Borja.

According to Rodrigues (1982), Aparicio Mariense da Silva attended the Itu Convention, as a member of the Republican propaganda in Rio Grande do Sul, and founded the Republican Club in Sao Borja. With each conquest, the Republicans became bolder, demonstrating their rebellion. In a motion in the Chamber registered in March 1883.

Rodrigues (1982) reports that there was no notable development in agriculture. However, all the rural properties had their own crops, which they produced for their own consumption. These included cassava for the production of flour, and sugar cane for the production of brandy, rapadura and molasses, the by-products of which were sold internally and exported to other municipalities. Products such as corn, wheat, beans, melons, watermelons and all kinds of vegetables were grown on the farms. There were some qualities of wheat, such as "Neapolitan", "Saracen", "Chilean Yellow" and other varieties, and "Canjica wheat". However, it also produced flour, with rolâo flour being very popular[3] . The orchards were abundant. Sâo Borja was called the "land of orange groves", which had been cultivated since Jesuit times. The production was so great that the people of Sâo Borja were nicknamed "papa laranja". The peach tree was cultivated on a large scale. The home industry

3 The coarser part of wheat flour that is separated from the ground wheat by means of a sieve and from which bread is made.

produced dried and syrup sweets, as well as the famous "origona cheeses[4] ". The quince was one of the most prized fruits because of its lifetime value.

Silveira (1979) writes of his arrival in Sâo Borja on horseback in 1886:

From afar we could see the orange trees, which we have been estimating for many years (without error or exaggeration) at around 5,000 trees, in almost every yard in the town and in the suburban houses.

The school, of which we still saw remnants with half a dozen cells converted into a barracks and military prison, the old houses of recollection, of the looms and workshops, were no longer to be seen.

Rodrigues (1982) reports that in 1887, after a population census, Sâo Borja received the honorary title of city.

Silveira (1979, p.251) "By law of December 21, 1887, the town of Sao Borja was elevated to city status".

Rillo (1982) reports that Sâo Borja stood out in terms of republican propaganda, having been one of the first municipalities in Rio Grande do Sul - the second, according to a good number of historians - to found its Republican Club, with Francisco Gonçalves Miranda as president and Aparicio Mariense da Silva as vice-president. This movement erupted in a major way in the so-called "plebiscite motion" of January 13, 1888. Two years before Princess Isabel freed the slaves, the Sao Borja City Council, by a majority of its members, had declared the municipality's slaves free.

Livestock farming had taken off during this period, with the establishment of ranches in various parts of the municipality's territory, and became the municipality's economic mainstay. Trade progressed, with the export of hides, jerky, rapaduras and cachaça to other regions via the River Uruguay, with semi-regular navigation at the time (RILLO, 1982).

In the 1893 revolution, the battle that hit Sao Borja was that of Carovy (RODRIGUES, 1982).

Rillo (1982) comments that in the 1893 revolution Sao Borja played a more prominent role, with many of its sons joining the "Northern Division", among them Colonel Manoel do Nascimento Vargas, as deputy commander, and the councillor and commissioned colonel Aparicio Mariense da Silva, as commander of the artillery corps.

1.2.1.4 Sao Borja and the 20th century

In the 1920s, there was the Alto Uruguai saladeiro, which processed beef and exported it, including canned beef, to the Brazilian market, either by land or by river (CSCJ, 1992).

Rillo (1982) comments that in the Revolution of 23, in Sao Borja, the federalists, called "maragatos", were commanded by Dr. Dinarte Rey Dornelles and Dr. Almeida Castro. They gathered their forces and took to the field. The Vargas family in Sao Borja, through their brothers Protàsio and Spartaco,

4 The peach is exposed to the sun to dehydrate. The peach is sweetened. The sweet peach paste, which has usually been braided or placed in a can, is added to the milk plus the rennet and placed in a sieve. The cheese is then matured (excess liquid is removed) and has the characteristic of being an unrefrigerated cheese.

organized a regiment of "provisionals" to defend the situation under the command of Colonel Deoclécio Motta.

According to O'donnell (1985), the political events that affected Sao Borja in 1923 were: Getúlio Vargas facilitating his re-election, and also the murder of doctor Benjamim Torres, on March 12, 1915; and the reinstatement of deputy superintendent Apparicio Dornelles Mariense by Decree No. 2188 of June 6, 1916. The 1924 revolution was part of a national revolution.

The 1930 Revolution, whose commander-in-chief was Dr. Getúlio Vargas, then Governor of the State. Some arrived in Rio de Janeiro, where, in an episode that became famous, they tied their horses to the Obelisk, in one of the central squares of the Capital of the Republic. In 32, during the Constitutionalist Revolution, Sao Borja organized the 14th Auxiliary Corps of the Military Brigade, under the command of Honorary Colonel Benjamim Vargas (Beijo) and Lieutenant Gregório Fortunato. The "14" took an active and even decisive part in the fighting in Sao Paulo. Some of the fighters from São Paulo went on to guard the Catete Palace and "personal guard" Getúlio Vargas (RILLO, 1982).

Teixeira (2011) reports on the invasion of Sao Tomé (Argentina) by Mr. Benjamim Vargas and members of the Provisional 14th. They only retreated at the end of the day, leaving behind a trail of blood and plunder.

Getúlio Dornelles Vargas[5] was born in Sao Borja on April 19, 1883. He was head of the provisional government of the 1930 Revolution, president elected by the Constituent Assembly on July 17, 1934, until the establishment of the Estado Novo dictatorship on November 10, 1937. He was deposed on October 29, 1945, returned to the presidency on January 31, 1951 by popular vote and on August 24, 1954 committed suicide while in office.

Rillo (1982) reports that agriculture took off in Sao Borja when Italian engineer Franco César Augusto Baglioni, hired by President Getúlio Vargas in 1935 and with the support of the state government, founded the Experimental Seed Station. He brought selected flax, wheat, alpiste, soybean, cotton and silkworm seeds from Cordoba.

Commenting on the profile of the local economy in the 20th century, Rodrigues (1982) observes that:

> [...] Sao Borja has always struggled and still struggles with the "distance" factor. Located in the extreme south of the country, achieving full socio-economic development becomes a difficult mission, since the municipality is far from consumer markets. Despite its rich soil, it was only from the 1930s onwards that agriculture began to grow, since until then agricultural production was only for domestic consumption. The arrival of settlers brought by some farmers contributed to this fact, developing flax cultivation, which was sold to Argentina, but it is worth remembering that the first experiments in agriculture were made around 1908, with the use of animal traction with disc plows in corn plantations. In times gone by, the town was called the breadbasket of the state of RS, where it was the largest producer of linseed and wheat. Among other crops that are very prominent in São Borjense's agricultural sector are rice, soybeans and sorghum, which have contributed to the establishment of a large number of mills and

5 http://www.culturabrasil.org/vargas.htm, accessed 20/10/2011.

cooperatives in the area (RODRIGUES, 1982, p. 149).

In February 1954, they founded the Regional Wool Cooperative "Missioneira" Ltda, with 74 members. On April 14, 1956, the cooperative was founded under the name "Cooperativa Agricola Samborjense Ltda", and at a general meeting on September 17, 1957, it was renamed "Cooperativa Triticola Samborjense Ltda" - COTRISAL. The founding group had 20 farming members. In 1958, the first headquarters was set up with the construction of buildings that today make up the seed department, with an initial capacity for 50,000 bags (FOLHA DE SÂO BORJA, 1981).

In 1960, Sâo Borja was the largest producer of flax in Brazil (Rillo, 1982).

Joao Belchior Marques Goulart[6] was born in Sâo Borja on March 1, 1918 at Fazenda Iguariaçâ. He died on December 6, 1976. Popularly known as **"Jango"**, he was a Brazilian politician and the 24th president of his country, from 1961 to 1964. Before that, he was also vice-president from 1956 to 1961.

Soybeans took off in a big way from 1965 onwards, taking turns as a summer crop with the land planted with winter wheat. After rice, it is currently the strongest crop economically in the municipality, always dependent on the greater or lesser regularity of rainfall (RILLO, 1982).

FAFISB and FESB[7] . Higher education began in the city on December 16, 1969, when the FAFISB - Faculdade de Filosofia Ciências e Letras de Sâo Borja was created. Until 1974, FAFISB remained an extension of UFSM (Federal University of Santa Maria), when it was recognized as an autonomous university and began to be maintained by FESB (Sâo Borja Educational Foundation),

URCAMP[8] , the community-based University of the Campanha Region, is established in Sao Borja. It was founded on February 16, 1989. In 1996, the Sao Borja Campus of the Regional University of Campanha was inaugurated.

The Cooperativa Triticola Samborjense (Cotrisal)[9] is managing to recover its finances and escape the liquidation process. A committee responsible for self-liquidation was set up in 1988, and in 1989, after analyzing the balance sheet, it became clear that Cotrisal was in the process of recovering, even though it was operating in the red.

The construction of the Sao Borja (BR) - Santo Tomé (Arg) International Bridge in December 1997. It is the only Unified Border Center in Latin America .[10]

1.2.1.5 Sao Borja and the 21st century

Pirahy II[11] . In 2000, the most modern rice processing plant in Latin America was inaugurated. In

6 Memorial CASA JOÂO GOULART- Sâo Borja - RS, 20/03/2012.
7 http://www.urcamp.tche.br/saoborja//, access:13 Jun. 2012.
8 http://www.urcamp.tche.br/saoborja/, access:13 Jun. 2012.
9 Zero Hour. Sunday, March 11, 1990.
10 http://www.mercovia-sa.com//, access:13 Jun. 2012.
11 http://www.pratofino.com.br/historia-pirahy/ access: 13 Jun. 2012.

2002, the company was renamed PIRAHY Alimentos. In 2007, the company opened its own parboiled rice plant with technologies and processes ahead of their time.

UERGS[12] , a state public university, is established in Sao Borja. Founded on July 10, 2001.

UNIPAMPA[13] , a federal public university, is established in Sao Borja. Founded by Law 11.640 of January 11, 2008. Admission from 2010 by the Unified Selection System (SISU).

IFF[14] , the Federal Institute of Farroupilha, a federal public institution, is established in Sao Borja. Founded on Bill 3775/2008. Classes start in 2010.

The biomass power plant[15] . The start-up of the thermoelectric plant powered by burning rice husks by investors from a German group - MPC Münchmeyer Petersen Real Estate Consulting GnbH, from the city of Hamburg, with an investment of R$ 55 million in Sao Borja. Rice husks are now used to generate energy and to make by-products from the ash. The biomass plant at full capacity will generate 12.3 megawatts of energy equivalent to supplying a city of 80,000 inhabitants. Inaugurated on May 25, 2012.

1.2.2 **Complementary clippings - Sao Borja**

This section will be sub-divided into: the indigenous influence; the Jesuit - Guarani influence: missionary period; the ethnic-political-cultural factors with a focus on Sao Borja; all of which belong to the historical context of the territory.

1.2.2.1 Indigenous influence

Freitas (1935) comments that the Guarani (warriors) were located between Paraguay and between the Paranà and Uruguay rivers, also occupying the left bank of the latter. They did not form an integrated nation, despite the curious phenomenon of speaking the same language, designated under the general name of "Guarani". Charruas - Dominated the region between the River Plate, Lagoa Mirim and Uruguay, citing Gay. Chased by the Spanish and the Paulistas, they settled on the northern bank of the Rio Negro and some of the hinterland of the Rio Grande. Guenoas, in the region between the headwaters of the Rio Negro and those of the Vacacai.

The Guarani had a complex social organization, in which leadership was obtained through prestige. The best warrior, the best hunter, the one with the best speech and the most generous was considered the chief" (SOARES and KLAMT, 2005, p.53).

1.2.2.2 Jesuit - Guarani influence: missionary period

Between 1568 and 1768, the Jesuit presence and their way of evangelizing changed the American continent, especially when they took the side of the natives, in the case of the Guaraní due to the

---

12 http://www.uergs.edu.br, access: 13 Jun. 2012.
13 http://www.porteiras.unipampa.edu.br/saoborja/, access: 13 Jun. 2012.
14 http://www.sb.iffarroupilha.edu.br, access: 13 Jun. 2012.
15 Folha de Sao Borja. Caderno Especial p.12, March 10, 2012.

Portuguese search for slaves for their plantations, especially in Sao Paulo, or the Spanish who sought out Indians for their orders, which were nothing different from enslavement (OLIVEIRA 2011a).

The Jesuit tried to adapt to the life of the Indian and to adapt the life of the Indian to his principles. I don't think there was any violence against the Indian in the sense of replacing one culture with another, but rather an accepted, communitarian approach, living together more freely, with more love and understanding. The result was a fraternal collectivism, of high justice, a communal organization of solidarity, in which the strongest did not swallow the weakest.

The other religiously-inspired socialist experiments attempted so far failed because they didn't develop practical forms of egalitarian coexistence. The Guarani republic escaped this defect, with its craftsmen, artisans, musicians and farmers on their way to the fields chanting and carrying images on their shoulders (OLIVEIRA, 2011a, p. 14 apud LUGON, 1979).

Freitas (1998) reports on the Jesuit-Guarani missions. They concretely and eloquently refuted the colonialist abuse that the Indians were incapable of sedentary life and higher forms of civilization, an argument used to enslave or exterminate them. In response to a colonialism based on landlordism and slavery, they created free, fraternal and egalitarian communities, using no other weapons than understanding and persuasion, in contrast to the massive and inhuman violence that marked the colonial enterprise. The reducción was a rigorously planned urban-rural unit. There were thirty or forty manufacturing workshops in each reducción, where blacksmiths, weavers, hatters, tanners, carpenters, potters, sculptors, painters, etc. worked. It was highly specialized and advanced. There were no idlers or parasites in the reductions.

This extraordinary experience of Jesuits and Guaranies, who were of an exceptional mutual solidarity, marks one of man's greatest attempts to promote the development of a primitive people, within the framework of social justice and respect..." (VENTURINI, 2009, p.61).

1.2.2.3 Ethnic - political - cultural factors with a focus on Sâo Borja

Langer (2005) reports that on the occasion of the administrative changes implemented in the reducciones with the expulsion of the Jesuits, the Guarani made the following statement: we are not slaves and we want to make it clear that we don't like the Spanish custom which wants everyone to look after themselves, instead of helping each other in their daily work. The Guarani were against colonial mercantilism. Private land ownership and individual work were non-existent categories in the socio-economic mentality of the pre-colonial Guarani.

Putnam (2002) argues that Latin Americans were hampered by the centralized authoritarianism, familism and clientelism they had inherited from medieval Spain, and the traditions of vertical dependence and exploitation were ignored.

Langer (2005) goes deeper into the individualist culture. He reports that for the Portuguese ranchers, linking the Guarani - Missioneiros to a supposed natural propensity for peasant skills was interesting for two reasons: firstly, because this ideological formulation legitimized the shameless land expropriation of the Indians, since, in the social logic of work at the time (and still today), a peasant was never and never needed to be a landowner. Secondly, because legally free labor was better suited to the reality of extensive cattle ranching, allowing for more efficient and less expensive exploitation of the workforce.

Torronteguy (1994) points out that the formation of the southern Rio Grande do Sul gaucho type can be traced back to the border struggles between Portugal and Spain. The bandeirantes, the tropeiros and smugglers from São Paulo and Laguna, the Jesuits, the Spanish adventurers, the Santafesinos, the indigenous tape, minuanos, charrua and guarani, the African slaves and the Portuguese soldiers and adventurers were the human types who, in a cultural and racial mix, made up a distinct human type for two centuries, from the first quarter of the 17th century to the first quarter of the 19th century: the gaûcho sulino. Their economic activities were restricted to cattle breeding, agriculture and smuggling.

Let's also agree that this lack of concern to delve into the truly regional phenomenon of the "gaûcho" - without forgetting the excellent efforts made by outstanding researchers from Rio Grande do Sul, Argentina and Uruguay - was a response to a general situation in which our respective countries were distanced from each other. Above all, political strategies for economic growth and integration into the world based on total national individualism had been consolidated since the end of the 19th century. The historiographical correlation of this "order" - a real "dis-order", in reality - of isolated South American countries, unaware of each other and, above all, full of national histories made **"to live alone",** as Alberto Methol Ferre clearly put it several years ago. Prior to the "gaucho homeland" (by its very nature mestiza), there existed a vast and clearly defined "missionary homeland".

(essentially indigenous) which was at the basis of the entire development of a vast territorial space which, when it fragmented with the Revolution (18101828), came to belong to the nascent states of Paraguay, Brazil, Argentina and Uruguay" (VENTURINI 2009, p. 62 and 63, apud FAVRE).

The author refers to the origin of private property, through the sesmaria, which established the rules of a distinctly pastoral economy and the cattle that populated the pastures, the coxilhas, the pampa and the fields of Rio Grande do Sul, always constituted the greatest wealth of the people. Laytano (1983, p.33): "The word estância means fixed dwelling, mansion where someone passes".

Torronteguy's book (1994) tries to explain a few points about the installation of capitalism in Rio Grande do Sul and the formation of the marginalized social segment. When the fields were fenced off, the dispossessed remained on the other side of the wire fence, outside the world of the landowners. The dispossessed helped build a class society geared towards the values imposed by those landowners, who at the time were involved in the export economy.

Venturini (2009) writes that agricultural production shifted from farmland to organized farming, including new crops brought from Europe such as cotton, wheat and grapevines. As well as the essential drink mate, with its health benefits. Cattle were the most important livestock introduced by the Jesuits, which guaranteed food for the missionary peoples and produced leather and tallow for export, making it the most important product in the Missions. Tobacco and sugar cane were also grown as agricultural products.

The different historical perspectives point to a so-called caudillesque regional culture, very close to the militaristic attitudes, political actions and social control practiced by Platine ranchers as agents of governments, and thus regulated by state regulations (FLÔRES, 2012).

Bakos (1982) comments on the importance of slavery in Rio Grande do Sul, as extensive agriculture and the salt meat trade developed in the final decades of the 18th century. A population survey from 1780 reveals the presence of slaves in the wheat-growing regions, in the charqueadas, in the first urban centers and in the first cattle ranches in Rio Grande do Sul.

Zarth (2006) comments that slavery in Sao Borja, in the missionary region, were areas where the presence of a free population dedicated to the extraction of yerba mate put downward pressure on the rates.

In the 19th century, the concentration of land, supply difficulties and slavery were the most important structural problems pointed out by political analysts and the government itself. The most effective alternative to these problems was the policy of immigration and colonization, combined with other projects to improve transport and technological innovation (ZARTH, 2006).

Multiculturalism was made up of other ethnicities, especially in urban areas through spontaneous immigration. On his military trip to Rio Grande do Sul, Gastao de Orléans (Count of Eu) commented that during the occupation by Paraguayan troops, several border towns such as Sâo Borja, Itaqui and Uruguaiana flew flags of different nationalities to protect themselves from looting and even death (CORRÊA, 2006).

Rillo (1982) reports that after the proclamation of the republic, in the last decade of the last century, the family of the German settler Pedro Gottfried arrived from Argentina. After 1910, on the initiative of the Vargas and Batista families, settlers with the surnames Frederich, Krug and Ramboner (a group of families still living today) came to settle in Sâo Borja, dedicating especially to growing corn and flax, on the properties of Protàsio Vargas (Estância Santos Reis) and Pedro Baptista, called "Rinçao dos Baptistas".

A key concept in this form of historical approach is undoubtedly "violence". The period between the Federalist Revolution of 1893 and the Revolution of 1923 is rich in episodes of both collective and individual violence. In this context, the power of the Borgist state and the PRR (Partido Republicano Riograndense - Riograndense Republican Party) was intertwined with the coronelist relations that significantly expanded the use of party violence and political co-optation in the formation of the complex theater of operations (COLUSSI, 2008).

In relation to prestige, the valorization of personality; the magazine[16] "o Rio Grande em Revista - [1928?]", with a photo registers in the localidade do Passo in Sao Borja a crowded "picnic" in offering to the then President of the Province, in figure 5.

---

16 O Rio Grande em Revista" magazine [1928?]. Apparicio S. Rillo Municipal Library. Sao Borja.

Figure 5 - Prestige in Sao Borja in Bairro do Passo

Source: O Rio Grande em Revista - [1928?], photo by the author.

1.3 CHARACTERIZING THE RURAL SURVEY ENVIRONMENT OF SMALL AND MEDIUM-SIZED PROPERTIES

This section will cover the characterization of the rural survey environment in Sao Borja-RS. Education is also a related subject.

The area of the municipality is 3610.0 km^2. The watershed is the Uruguay River and the Butui, Piratinim and Icamaqua rivers. The subsoil contains the Guarani aquifer reservoir. The terrain is predominantly 0-2% slope, covering 72% of the territory. The municipality's economic profile is based on the primary sector with rural agricultural activities, which are the mainstay of the local economy, and livestock farming (LAGEOLAM, 2007).

The *city census (2010)* contains information about the municipality:

Table 3- Sao Borja City Census (2010) rural, urban and total population

Rural population	Urban Population	Total Population
6.533	55.138	61.671
10,6 %	89,4 %	100%

Source: IBGE Cidades (2010) Code 431.800. Adapted by the author.

Table 3 shows the small rural population in the municipality of Sao Borja.

Table 4- Rural resident population, age group and percentage

Resident population - total - rural	6.533	people
Resident population - rural - age groups From 0 to 5 years old	8,2	%
Resident population - rural - age groups From 6 to 14 years	16,0	%

old		
Resident population - rural - age groups 15 to 24 years old	14,6	%
Resident population - rural - age groups 25 to 39 years old	20,5	%
Resident population - rural - age groups 40 to 59 years old	27,8	%
Resident population - rural - age groups Aged 60 and over	12,8	%

Source: IBGE Cidades (2010) Code 431.800. Adapted by the author.

This table 4, has a higher percentage in the rural resident population stratum

in the 40 to 59 age group. There has also been a sharp decrease in the population aged over 60. It can also be seen that there is a decrease in percentage corresponding to the 15 to 24 age group in the rural population, indicating the rural exodus, and in percentage terms there is an insufficient population of 0 to 5 year olds, with fewer births.

According to information provided by the 35ª Coordenadoria de Educaçao - Sao Borja - RS, at the Franco Baglioni and Militina Pereira Alvarez State Elementary Schools and the Timbaùva State Middle School, there are data for the years 2000, 2005, 2010 and 2012 on the number of students initially enrolled in the respective rural schools in the municipality.

Graph 1 - Number of pupils in rural schools from 2000 to 2012 in the municipality of Sao Borja - RS

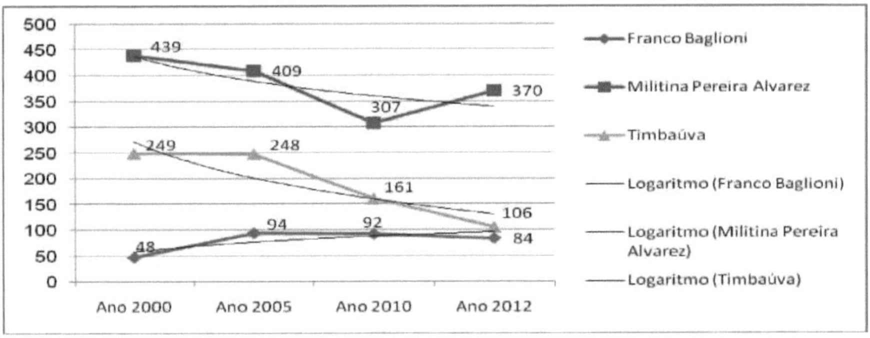

Source: School Census (2000, 2005, 2010) and PROCERGS 2012, adapted by the author.

Graph 1 shows trends indicating a decline in pupils at two rural schools and one with stability. The Franco Baglioni school received students from rural settlers from 2005 onwards, stabilizing, but with a downward trend thereafter.

2 METHODOLOGY

The study adopted a qualitative approach to identify the main factors involved in the problem and possible alternatives for intervention. According to Gil (1991), this type of research favors the consideration of historical factors, which are fundamental for understanding social structures. In this direction, procedures were set up to understand conceptions, perceptions, behaviors, feelings, meanings and changes inherent in the context of the phenomenon being researched. The quantitative approach was used to translate opinions and information into numbers, classify them and analyze them, using statistical techniques (SILVA, 2004).

As far as its objectives are concerned, from the perspective of Cooper and Schindler (2001), the study is characterized as exploratory and descriptive. It was exploratory in that it aimed to get to know a universe that had been insufficiently researched with regard to the analysis of determining factors, their correlation with the backwardness of the region surveyed and the proposal of "new" sustainable development strategies. It is also descriptive, as it seeks to provide in-depth information and data on these aspects. According to Gil (1994), exploratory research has as its main objective the improvement of ideas or the discovery of intuitions and descriptive research makes it possible to survey the opinions, attitudes and beliefs of a population.

As for the technical procedures, based on Silva (2004), the following were used:

Bibliographic research, based on published material, consisting mainly of books, journal articles and material available on the Internet;

Documentary research, based on materials that have not been analyzed;

The *survey*, involving direct questioning of producers characterized as "family farming", whose behavior was intended to be known;

The Case Study, involving the in-depth and exhaustive evaluation of the specifics of a singular problem, in order to obtain a broad and detailed knowledge of it;

Ex-Post-Facto Research, since the study dealt with facts that had already happened, even though the situation analyzed was contemporary; and

Action Research, since the study was conceived and carried out in close association with the proposal of alternatives for solving a collective problem, in which the researcher and the members of the problem situation were involved in a participatory way.

The data and information needed to carry out the study was obtained using four basic tools: o "Survey of the theoretical framework on endogenous, exogenous and neoendogenous development"; o "Survey of the history of the municipality of Sao Borja-RS"; o "Survey of the socio-productive situation of small and medium-sized rural properties in the municipality of Sao Borja-RS"; o "Survey of human and social capacity in the development of the municipality of Sao Borja-RS", with the aim of covering cognitive-behavioral aspects not covered in the previous survey;

Details of the method used in each are presented below.

2.1 SURVEY OF THE THEORETICAL FRAMEWORK ON ENDOGENOUS, EXOGENOUS AND NEOENDOGENOUS DEVELOPMENT

The information and data researched in this section came basically from secondary sources, especially books, dissertations, theses and scientific articles.

To interpret the data, the method of content analysis was used, using the technique of thematic analysis. According to Bardin (2011), this procedure involves evaluating one or more themes or items of meaning, previously coded.

The aim of this task was to highlight important information to help with the questions included in the surveys of the socio-productive situation and the definition of the possibilities for neo-endogenous development.

2.2 SURVEY OF THE HISTORY OF THE MUNICIPALITY OF SAO BORJA - RS

Roesch (2010) *apud* Patmore (1998) considers that the historical method allows theories to be generated and tested from a long-term perspective, enabling researchers to develop dynamic approaches in contexts of economic, political and social change. On its own, this method is insufficient for investigating the degree of association and causality between variables. However, it is very useful in defining and justifying the important aspects to be investigated.

Costa; Barros; Martins (2010) comment on the integrationist position that can also be identified in the *management history* approach. The idea is that historical analysis does not replace existing theories, but enriches the understanding of the current context through connections with humanistic disciplines. Adopting the integrationist position means recognizing that past events act in the present, being useful for identifying and choosing current opportunities.

A reasonable portion of the historical reference presented in section 1.2 had been researched by the author in books, works and documents on the Jesuit-Guarani, Spanish and Portuguese, and later Brazilian missionary period and on the 330-year history of São Borja. In 2012, the author even took part in the supporting committee for the event commemorating the 330th anniversary of São Borja, where he gave a lecture on the town of São Francisco de Borja das Missoes from 1801 to 1834: its Luso-Brazilian formation, invasions and ethnic, political and cultural factors (SoUTo, 2012).

The data collected in the historical research came from documentary sources, consisting mainly of manuscripts, printed matter and photographs; and secondary sources, such as books, theses, magazines and newspapers.

To interpret the data, the method of content analysis was used, using the technique of thematic analysis. According to Bardin (2011), this procedure involves evaluating one or more themes or items of meaning, previously coded.

The aim of this task was to highlight important information to support the preparation of questions included in the socio-productive profile surveys; and the proposal of intervention alternatives - with a focus on identifying the main past social, political and economic factors, presumably determining the recent situation. All in line with the specific objectives of the study and the underlying theoretical and empirical foundation.

2.3 SURVEY OF THE SOCIO-PRODUCTIVE SITUATION OF SMALL AND MEDIUM-SIZED RURAL PROPERTIES IN THE MUNICIPALITY OF SAO BORJA - RS

Properties smaller than 100 ha account for almost 73% of the total calculated by the 2006 Agricultural Census (Table 5).

Table 5 - Number of rural properties in the municipality of São Borja-RS, by area strata

Strata	**N° Properties**
More than 0 to less than 0.1 hà	3
From 0.1 to less than 0.2 hours	0
From 0.2 to less than 0.5 hours	1
From 0.5 to less than 1 hour	1
From 1 to less than 2 hours	36
From 2 to less than 3 hours	43
From 3 to less than 4 hours	29
From 4 to less than 5 hours	30
From 5 to less than 10 ha	89
From 10 to less than 20 ha	168
From 20 to less than 50 ha	179
From 50 to less than 100 ha	111
From 100 to less than 200 ha	108
From 200 to less than 500 ha	151

Source: Brazilian Institute of Geography and Statistics, Agricultural Census (2006).

In addition, the vast majority of these properties can be defined as "family farms", according to the size criteria contained in the pronaf Support Declaration (DAp). Given these circumstances, and especially the fact that these establishments concentrate the most impoverished and needy strata, it was decided to consider the study population to be 690, i.e. those with areas of less than 100 ha.

To determine the sample size, the desired confidence level and the accepted margin of error were taken into account. The formula used to determine the sample size, taken from Rea & Parker (2000), is as follows

$$n = \frac{Z^2[p(1-p)]N}{Z^2[p(1-p)]+(N-1)C^2} \qquad (1)$$

where C is the precision or maximum admissible error in terms of proportions; Z is the confidence level in standard deviation units (desired range); p is the proportion of the universe; N the number of elements in the population; and n is the number of elements in the sample.

For the calculation, we defined a 95% confidence interval and a maximum margin of error of ± 5%, solving equation (1) for C = 5%, Z = 1.96, P = 50% (a conservative percentage which results in the highest possible response rate) and N = 690. The result for n is 248.

A similar result was also found for the calculation according to the methodology recommended by SEBRAE (2005) for carrying out market research.

The stratification of properties, for the purposes of sampling, satisfactorily met the proportion of 35.9% (248/690). The lower percentages in two strata and the higher in one do not compromise the representativeness of the sample as a whole. The largest population stratum was covered by 41% (Table 6).

Table 6 - Determining the sample size of rural properties

Total area groups	**Population (A)**	**Sample (B)***	**B/A (%)**
From 0 to less than 10 hours	232	74	32
From 10 to less than 50 hours	347	141	41
From 50 to less than 100	111	33	30
Total	-	248	-

Source: Author's elaboration.

* Sample error=+/- 5% [Equation (1)]; and Sample error=+/- 5% with *Split* 50/50 (SEBRAE, 2005).

To support the definition of the questions in the data collection form, in addition to the results of the lexical analysis of the historical findings in section 1.2, a content analysis was carried out of the other theoretical and empirical references compiled in sections 1.1 and 1.3.

The aim of this study was to provide information revealing presumed strengths, opportunities, weaknesses and threats, as well as feasible alternatives for intervention - with a focus on identifying the main social, political and economic factors supposedly determining the recent situation and its prospects for development. All in line with the specific objectives of the study and the underlying theoretical and empirical foundation.

In order to determine gross monthly income (an item on the form), data on reported sales, which could be quantified, and socio-economic classification were used. In some cases, the author/researcher used criteria that he had defined, putting himself in the position of an expert. Firstly, the items on agricultural, cattle, sheep, pig and honey sales and the number of pensions converted into values were quantified. The value of the products sold was calculated from the quantities and average prices based on the data provided by Emater/Sâo Borja for the period September-December 2011, and those taken from a monthly edition of the newspaper Zero Hora,

Campo e Lavoura supplement for January-July 2012. With regard to sales of subsistence crops, vegetables, poultry and eggs, fish, fruit and other crops, the amount sold was not quantified, given the precariousness of this information.

Another criterion for estimating gross monthly income was socio-economic classification, following the procedure adopted by the Brazilian Association of Market Research Institutes - ABIPEME (2008). In this task, the author once again acted as an expert, defining the classification range, in the absence of data on commercialization, questioning and tabulating information on the ownership of color televisions, video cassettes or DVDs, radios, bathrooms, cars, monthly maids, washing machines, refrigerators and freezers.

Despite the subjectivity involved in the author's meddling, the effort was valid, since it was not a question of determining rural costs and expenses, but only of estimating gross monthly income, for the purposes of classifying each property in the socio-economic classification adopted.

The data collection form used closed (yes/no), open and multiple choice questions, with and without the option of a Likert attitude scale (Appendix A).

The field survey was carried out between Sept./2011 and July./2012, and the research was registered in the Postgraduate and Extension Management Support System (SAPE) - Urcamp/RS, under No. 612/2011.

The data was collected through random visits to farmers on the properties along the roads. To do this, we used maps provided by the Municipal Planning Department, telephone directories of farmers, as well as schedules obtained at events, rural festivals, rural masses, etc.

For the qualitative analysis of the data, the method of content analysis was used, using the technique of lexical analysis. This approach consists of computing the total number of words present or "occurrences"; or the total number of different words or "vocabulary" and the ratio of occurrences to vocabulary, contained in the answers to the open questions (BARDIN, 2011).

In the quantitative analysis of the data, the statistical method was used, using the techniques of relative participation (%), tabulations (simple, cross and multiple) and correlations.

The SPHINX® *software*, currently used by more than 200 Brazilian research institutes and universities, including FGV, USP, UFRGS, UFSC, etc., was used for qualitative and quantitative processing.

It is also used to present tables in Excel and the free Freemind software for branched graphic formatting.

2.4 SURVEY OF HUMAN AND SOCIAL CAPACITY IN LOCAL DEVELOPMENT - SÂO BORJA-RS

The same methodology used by Oliveira (2011b) was used to verify the opinion of urban and rural residents about the current situation in terms of their relationships of trust, leadership initiatives, entrepreneurship and level of schooling. Although the target population is urban and rural, the respective indices were measured exclusively for complementary purposes.

Table 7 was used to define the sample size. For a population of 61,671 inhabitants in the municipality of Sao Borja (IBGE 2010), a sample of 100 people was selected to be interviewed using the form with quantitative and qualitative questions in Annex A. In order to ascertain other qualitative opinions, a non-probabilistic intentional sample was used, in which 5 people were heard through a recorded interview. The interviewees were the mayor of the municipality for the 2009-2012 administration, a technified small rural producer who specializes in irrigation, a retired rural worker who was a farmhand for former President Jango Goulart and witnessed visits by former President Vargas to the farm where he worked, a settled rural producer (community leader), a priest from the Parish of Our Lady of the Conception - Passo district, who works with the solidarity economy and family farming.

For data collection, the sample stratification followed the mapping of the Health Department of the Municipality of Sao Borja - PSFs (Family Health Program).

Table 7- Determining the amplitude of a population sample

Population size (Universe)	Sample size with margins of error	
	+ - 5 %	+ - 10 %
6.000	375	98
15.000	390	99
100.000	398	100
More than 100,000	400	100

Source: Prepared by OLIVEIRA (2011b) (apud GIL, p.124, 2002).

For the estimates of education and schooling, the information was obtained from IBGE (2012).

The data was tabulated and analyzed using SPHINX *software,* and the qualitative data was subjected to content analysis with the help of lexical analysis.

3. RESULTS AND ANALYSIS

This chapter presents the main results and the respective analyses carried out on the basis of the data and information obtained in the investigations into the theoretical-empirical-historical framework, the socio-productive situation, and human and social capacity in the region surveyed.

3.1 ENDOGENOUS, EXOGENOUS AND NEOENDOGENOUS DEVELOPMENT

Using the content and citation analysis carried out on the texts consulted and summarized in subsections 1.1.1 and 1.1.2 of Chapter 1 (Theoretical and Empirical Framework), we used the methodology defined in subsection 2.1 of Chapter 2, where the content was classified into categories, subcategories and citation sources (Table 8).

Table 8 - Identification of the main aspects in the categories and subcategories to be considered in neo-endogenous development projects

Categories	Sub-categories	Sources of citations
I) Capital Human/Cultural	**1.1) Individuals and families;**	Bordieu (2000), Chiavenato (2000,2004), Dallabrida (2011), Fukuyama (1996), Higachi;Clemente (2000), Haucap;Wey (2000), Lucas (1988), Paiva (2001), Schultz (1962);
	1.2)Intellect-Capabilities: • Conceptual Skills - Ethics; • Human Skills - Communication, Motivation, Leadership, Teamwork, Empowerment; • Technical Skills - (Having the Knowledge) Knowledge, doing concrete and practical things; • Managerial Skills - Planning, Organizing, Directing, Controlling; • Attitude- (Wanting to do) - Entrepreneurial flair, risk-taking, innovation;	Chiavenato (2004,2007), Friedmann (1966), Guha (1981), Hirschman (1958;1961;1977), Katz;Khan (1987), Marshall (1982), Perroux (1955;1959;1967; 1977), Radomsky (2011); Rambo;Costa (2011), Schumpeter (1961,1985), Silva (2004), Stoner; Freeman (1995), Wilkinson (2008);
	I.3) Continuing education;	Abascal;Bilbao (2010),Baquero (2008), Barros; Mendonça (1997),Bordieu (1998), Breitenback;Brandâo (2009), Eco (1998), Freire (1985),Gadotti (2002),Jacobi (2003), Keil;Barquero (2008), Marshall (1982), Milinski;Campanhol;Silva (2008), Oliveira (2011), Rostow (1961), Rotta (2007),Sen (2000),Valdiviedo (2003).
II) Share Capital	**II.1) Actors in the community;**	Abramovay (1998), Allebrandt, Büttenbender, Siedenberg (2010) Alves (2008), Bikholzer (2005), Brum (2005), Clemente; Higachi (2000), Dallabrida (2011), Gazolla;Pelegrini (2011), Giddens (1984), Fukuyama (1996), Long (2006), Long;Ploeg (2011), Mior (2005),
Categories	**Sub-categories**	**Sources of citations**
		Schneider (1999), Yunus (2000).

	11.2) Cultural diversity: • Ethnicities; • Gender; • Local knowledge;	Cremonese (2011), Long;Ploeg (2011), Radomsky (2011); Rotta (2007,2011), Sauer; Correa (2011), Sen (2ND), Silva (2009), Schneider; Gazolla (2011), Souza;Souza (2011), Veiga (2005);
	II.3) Trust;	Baquero (2003), Fukuyama (1996), Kliksberg (2000), Pase; Santos (2008), Putnam (1996; 2002);
	II.4) Networks and Partnerships;	Abascol; Bilbao (2010);Alves (2008), Araùjo (2005),Atria (2003),Birkholzer (2005) Castells(1999;2002),Julien(2010), Mance(2001);Milinski;Campanhol;Silva (2008), Radomsky (2011); Souza;Souza (2011), Putnam (2002),Vaz (2010)
	II.5) Entrepreneurship (individual and collective);	Abramovvay (1998),Ansoff (1991), Antunes; Ries(2001),Cella; Peres(2002), Chiavenato (2008),Gimenez;Tóffolo (2005), Hisrish;Peters (2006), Julien (2010) Longenecker;Moore; Petty (1998), Machado et.al(2000),Schumpeter (1985) Shapero (1975),Zuin;Queiroz (2006);
	II.6) Associativism and Cooperativism;	Albano (2001), Araùjo (2005), Baquero (2008), Büttenbender (2010), Campos (1998), Crùzio (1997), Duarte (1986), Franco (1985), Moura (1964), Oliveira (2011), Perius (1983), Riesgo;Brezolin (1988), Silva et. al (2008);
III) Political Capital	**III.1) Power structure;**	Allebrandt (2008), Alves (2008), Baquero (2007), Baquero, Cremonese (2008), Baran (1977), Bendix (1996), Bordieu (1998;2000;2003), Castells (1999),Clemente (2000), Daniel (1982), Ellis; Biggs (2001), Esteves (2009), Gonzâlez (2007;2008), Jaguaribe (1972), Kebauy; Assumpçâo (2011), Schugurensky (2003), Veiga (2001), Weber (1984), Weyh (2008);
	III.2) Political leadership;	Gadotti (2002), Hannes (2005), Morgan (2006) Schumpeter (1985), Wagner III; Hollenbeck (2002);
	III.3) Political participation	Bordieu (1998;2000;2003), Freire (1999), Jaguaribe (1972), Julien (2010), Schugurensky (2003), Weyh (2008);
	IV.4) Empowerment;	Baquero (2005), Durston (2001;2003), Freire (1985), Gadotti (2002), Gohn (2004), Hannes (2005), Pase (2001), Pase,Santos (2008), Pereira (2006), Schumpeter (1985), Rambo;Costa (2011), Wallerstein;Bernstein (1994), MDA/DESER (2008);
IV) Financial Resources	**IV.1) External resources:** • Non-repayable; • Availability of credit; **IV.2) Internal resources/savings**;	Alves (2008), Barea; Miorin (2008), Boisier (2007), Breitenbach; Brandao (2009), Cardoso; Falleto (1985), Dallabrida (2011), Furtado (1970), Keynes (1985), Lucas (1988), Prebisch (1961), Rostow (1961), Schumpeter (1961;1985), Solow (1956), Swan (1956), Veiga (2001); - Birkholzer (2005), Rostow (1961), Putnam (2002);
Categories	**Sub-categories**	**Sources of citations**
V) Physical and Material Resources	**V.1) Earth;** **V.2) Water;** **V.3) Electricity;**	Antunes; Ries (2001), Clark (1940), Dallabrida (2011), Ferreira (1989), Hirschman (1958;1961;1977), Isard (1956,1972), Marshall (1982), Myrdal (1962,1968), North (1955, 1977a,1977b), Perroux (1955;1959;1967;1977), Porto (2007) Rosa (2011), Rosenstein-Rodan (1963), Schultz

	V.4) Inputs; **V.5) Roads;** **V.6) Livestock;** **V.7) Infrastructure;** **V.8) Machinery and equipment;**	(1965;1975), Schumpeter (1961);
VI) Resources from Operating Activities	**VI.1) Products;** **VI.2) Processes;** **VI.3) Diversification;**	Abramovay (1998), Abascal;Bilbao (2010), Antunes;Ries(2001), Boisier (2005), Castells (1999;2002), Cella (2002), Clark (1940), Franck (1980), Perroux (1955;1959;1967;1977), Rosa (2011), Rosenstein-Rodan (1963), Rostow (1961), Schneider (2003;2006), Schumpeter (1961;1985), Triches;Froehlich;Schneider (2011), Wilkinson (2008), Vaz (2010);
VII) Marketing resources	**VII.1) Local buyers;** **VII.2) External buyers;** **VII.3) Contracts;** **VII.4) Consumer habits;**	Abramovay (1998), Castells (2002), Dallabrida (2011), Gazzola; Pelegrni (2011), Harris (1954), Hofling (2001), Julien (2010), Keynes (1985), Myrdal (1962,1968), North (1955;1977a; 1977b), Schneider (2003), Schumpeter (1961), Schultz (2011), Schwartzman (1977), Souza;Souza (2011), Triches;Froehlich;Schneider (2011), Valle (2010), Wilkinson (2008);
VIII) Institutional Resources	**VIII.1) Technical assistance and rural extension;** **VIII.2) Educational institutions (universities and federal institutes);** **VIII.3) Government Organizations;** **VIII.4) Non-Governmental Organizations;**	Abascal;Bilbao (2010), Capoal;Costabeber(2001), Carmo (1998), Cella (2002), Dallabrida (2011),Marin,Trentin (2010), Esteves (2009),Ferreira (1989), Furtado (1985,2004), MDA/DESER(2008), Milinski;Campanhol; Picolotto (2011); Rangel (1957; 1986;1990;1992),Sachs (1998), Silva (2008), Schultz (1965;1975), Souza (1983), Triches;Froehlich;Schneider (2011), , Valle (2010),Zonta (2012);
IX) Logistical resources	**IX.1) Storage and stockpiling;** **IX.2) Internal transportation;** **IX.3) External transportation;**	Dallabrida (2011),Isard (1956;1972),Marshall (1982), Triches;Froehlich;Schneider (2011), Wilkinson (2008)
X) Technological/Informational Resources	**X.1) Technology;** **X.2) Information**	Castells(1999,2002),Dallabrida (2011), Fukuyama(1996),Higachi;Clemente (2000), North (1955; 1977a;1977b), Padovese (2009), Rosenstein-Rodan(1963), Rostow (1961), Schultz (1965;1975),Schwartz (2000),Vaz (2010)

Source: Prepared by the author based on Bardin (2011); Meireles and Cendón (2010); and sub-sections 1.1.1 and 1.1.2.

Table 8 summarizes the main categories, subcategories and sources of citations from the reference framework and other works consulted. The categories are human/cultural capital, social capital, political capital, financial resources, physical and material resources, operational resources, marketing resources, institutional resources, logistical resources and technological/informational resources. The aforementioned categories were brought together and represented by the content analysis carried out using this methodology. These categories and subcategories will be listed in section 3.5.

3.2 HISTORY OF THE MUNICIPALITY OF SAO BORJA-RS

In order to facilitate a global view of the temporal evolution of the historical records of the municipality of Sao Borja-RS, Table 9 summarizes the main aspects relating to settlement (ethnic groups), primary products and production systems at different times.

Table 9 - Historical-sociological analysis of Sao Borja: settlement (ethnic groups), primary products and production systems

Season	Population (ethnic groups)	Primary products	Production systems
Century XVII	Indians, Jesuits.	• native fruit pickers; • hunting; • fishing; • subsistence crops.	• native forests; • native fields; • areas of the Jesuit reductions.
Century XVIII	Indians, Jesuits, Spaniards.	• yerba mate; • cattle, mules, horses; • subsistence crops.	• areas of the Jesuit resorts; • cowsheds in the mission region; • tropeirismo, mainly of mules to Sao Paulo and Minas Gerais.
Century XIX	Indians, Spaniards, Portuguese, Africans, Brazilians, immigrants: Prussians-Germans.	• cattle, mules, horses, sheep; • orange trees, peach trees; • cassava, wheat.	• sesmarias, Luso-Brazilian estancias and chàcaras; • tropeirismo for the beef jerky in Pelotas; • the beginning of agriculture in the municipality; • cassava and wheat flour factories.
Century XX	Missionaries, Brazilians, Afro-descendants, immigrants of German, Russian, Italian, Polish "whitening" and Arab descent.	• cattle, horses, sheep; • wheat, flax, linseed; • soy, rice, corn.	• resorts and chalets; • charqueada Alto Uruguai (Sao Borja) of cattle and canned goods; • beef packing plants outside the municipality; • agricultural cooperatives: beginnings, heyday and decline; • rice mills; • properties: small (the beginning of the term family farming), medium and large private properties.
Century XXI	ethnic mix	• cattle, horses, sheep; • rice, soy, wheat, corn;	• rice mills; • properties: small (family farming),medium

		• rice husk energy.	large private properties and large corporate properties; • agro-industries of beneficiation and aggregate products; • Refrigeration outside the municipality.

Source: historical research on the 330th anniversary of Sao Borja, adapted by the author

In the 17th century, with the second settlement of the Jesuit reductions, the territory was populated by different indigenous ethnic groups. These had different traditions. The Guarani tradition maintained small gardens with different crops, fished, hunted and gathered wild fruit. They were free and migrated. The cultural-political system used cacicado and prestige. With the arrival of the missionaries, some accepted catechization with the benefits of having regular food, but they had to comply with the organization and systems of the reduction, as well as converting to the Catholic religion. Some Indians didn't accept this and went to live in the forests. At this time, the Jesuits introduced collective farming and family farming. The priests also introduced cattle as a source of food for the reductions, as well as horses and mules, and the indigenous custom of drinking yerba mate. Fruit growing was introduced and subsistence farming.

In the 18th century, the Jesuit reducciones and estancias grew and developed to the point of arousing the greed of the Iberian kingdoms, the colonialists and individualists. They waged war against the Guarani Indians, forcing the mercantilism of the time. Yerba mate was the target of mercantilist greed. In 1756, the trial took place in San Borja of the indigenous leaders who had fought against the Iberian alliance. There were several treaties in which the missionary region of the Seven Peoples of the Missions was Spanish and then Portuguese. And again it returned to the Spanish crown. In this context was the town of San Borja (the Borjistas). With Spanish rule, the model of collective labor in the fields was extinguished when the Jesuits were expelled in 1759. The indigenous people were exploited, expropriated because they had no formal education and became the dispossessed. Along with the increase in cattle, there was an increase in horses and the crossbreeding of donkeys with mares led to the creation of mules. These were of economic interest due to their ability to transport cattle to the São Paulo and Minas Gerais markets. Cattle were slaughtered and little was used: leather, horns, tallow and tongue; and the meat was only used to satisfy hunger.

The 19th century began and the fighting intensified, but the Guarani joined the Portuguese to the detriment of the Spanish in 1801, who took the town with little resistance. Portuguese rule with its individualism began. The Jesuit-Guarani period was treated with oblivion, as it was linked to Spanish rule. Sesmarias and then large estancias were granted to Portuguese soldiers and/or people of means. The indigenous people were dominated by the power of arms. The latter were used to manage the cattle, as they were already adapted to farming. A providential domination, of subject (man, white and European) with objects (Indians, illiterate, passive and manipulable) because all that was needed was a piece of meat, yerba mate, a hut and a companion to consolidate the domination of the Indians. Châcaras and postmen's dwellings appeared on the large estancias. They were the peons and soldiers, the gaûchos missioneros (gaudérios) at the service of those with power and possessions. The period of the military commanders began. There was the first Caudillesque invasion in 1816 (Andresito Artigas, an eastern Platino-Guarani), so there was a lot of death and resistance from the Portuguese-Guarani in the town of São Borja. It forced the Portuguese to turn the town of San Francisco de Borja into a military garrison, and with this the young indigenous people were assigned to military practices instead of agriculture.

In 1821, Saint Hilaire spent February and March in the town of São Borja. He describes the misery of the Indians, noting how they had become paupers and wanderers, the neglect of the reducción and the military focus. In 1822, the Brazilian Empire was established, so Sâo Borja was a Gaúcha missionary and a Brazilian. In 1828, the caudillo Fructuoso Rivera (Argentine-Guarani) invaded again, causing a lot of damage and looting. For eight months, Fructuoso created the Eastern Province of Uruguay, whose provincial capital was San Borja. In 1831, the National Guards were organized.

In 1834, the town became a village. In the intervals between the "skirmishes", the remaining cattle reproduced themselves and there was tropeirismo, but for the charqueadas of Pelotas.There was the Farroupilha revolution where there was little fighting in Sâo Borja. In 1843, the Empire was granted possession of the lands in the former estancias of the Indians of the Eastern Missions. 1850 saw the entry into force of the Land Law, which allowed land to be sold. Slavery in Sâo Borja in 1859 numbered 1,240 slaves In 1865 there was a Paraguayan invasion, with fighting, deaths, heroism and looting. After a long period of reconstruction, there was a republican propaganda movement against slavery and the monarchy in 1874. This was a period of colonelism and large estanciers, where paternalism-clientelism-familism was at a standstill. The arrival of Prussian-German settlers was arranged and agriculture, especially wheat and flax, was given a boost in 1885. Sao Borja was known as the "land of orange groves" and the people of Sao Borja as the "orange pope". In 1887, Sao Borja was elevated to the status of a city with only 6 slaves. Troops of cattle continued to the charqueadas in Pelotas. In 1888, the City Council approved the Republican Plebiscite Motion. There were artisanal manioc flour factories and wheat mills. The Republic is proclaimed in 1889. The revolution of 1893 took place, consolidating the Vargas political command.

In the 20th century, improved European and Zebu breeds were introduced to livestock farming. German and white Russian settlers also arrived on the estates of famous ranchers. In 1911, the Charqueada Alto-Uruguay partnership was formalized, with Uruguayan-English capital and Brazilian partners. A strong period of politics, abuse and violence related to the PRR (Partido Republicano Riograndense). Start of the railroad in 1913, the B.G.S. (Brazilian Great Southerns Railways Extensions Ltd.), linking Sao Borja. Completion of the Alto-Uruguay Charqueada in 1914 due to the war and the change in competitors' technology. In 1920, the Wilson-Armour meatpacking plant began in Santana do Livramento. In 1923 the revolution against Borges de Medeiros took place and the situation won, coronelismo. 1928 Getúlio Dornelles Vargas is proclaimed President of the State of Rio Grande do Sul. The 1930 national revolution takes place, in which São Borjenses participate directly and President Getúlio Dornelles Vargas is put in power. In 1933, Santo Tomé (Arg.) was invaded by Sao-Borjans. The policy of domination by the large estancias continues, with châcaras and posteiros remaining as aggregates. 1935 sees the founding of the Experimental Station of Seeds. Sao Borja was considered the land of wheat, flax and linseed production.

The second cooperative model was promoted by the government from 1950 onwards. Italian descendants move to Sao Borja. Jango comes to power, but is exiled in 1964. The 1970s saw the consolidation of cooperativism. In the 1980s and 1990s, cooperativism went into crisis. Then came the cycle of soy, wheat and rice. Rice grew stronger and rice mills were set up by entrepreneurs of Italian and German origin. The term "family farming" was adopted in Brazil under Fernando Henrique Cardoso. Under the Lula government, programs aimed at family farming were planned and implemented.

At the end of the 20th century and the beginning of the 21st century, agricultural cooperativism in São Borja has fallen into disrepute. The cooperative principles need to be strengthened, such as loyalty to the statutes, effective participation, communication, professional management, alternating leadership and self-management. At the end of the 20th and 21st centuries, we saw the decline of traditional families in livestock farming, not least because of family inventories, to the detriment of the growth and increased importance of individual entrepreneurs of Italian and German origin, preferably. These predominate in Sâo Borja's economic and political power, both in agriculture, livestock, commerce, agro-industry and services in the municipality.

The new historical-rural profiles in the 21st century also come from rural entrepreneurs who have acquired land at lower prices in São Borja than their places of origin, and invest in the rural area of the municipality. Acquiring from landowners with a traditional, unproductive and inefficient profile; those who sell their land and assets because they are not competitive with the new times of globalization or because of family breakdowns fragmenting the old rural areas. This segment includes liberal professionals (doctors, lawyers, etc.) who, with surplus income, invest in land to increase their assets or for rural recreation.

Also in this historical development, the arrival of private companies: Gaucho, national and foreign, in all economic-productive segments, is representative and growing in the municipality. These private companies are strongly active in the municipality's agribusiness segment, especially in the wake of the decline of cooperatives, where they have seen business opportunities in land, processing/trading of commodities and energy. They are professionalized with economic power and are technologizing livestock, agriculture and forestry; however, they are minimizing manpower. Therefore, in business opportunities, they acquire rural properties from small, medium and large landowners; or by exchanging debts with the companies themselves, increasing their economy of scale. They have already changed the stratified rural profile with large tracts of land in Sao Borja.

The content and citation analyses carried out on the texts consulted and summarized in Section 1.2 of Chapter 1 (Theoretical and Empirical Framework) were conducted with the primary purpose of identifying past factors, presumably determining and conditioning rural development in the municipality of Sao Borja-RS. Using the methodology defined in Section 2.1 of Chapter 2, the contents and citations were classified into categories, subcategories and citation sources (Table 10).

Table 10 - Identification of past factors presumably determining and conditioning rural development in the municipality of Sao Borja-RS

Categories	Sub-categories	Sources of citations
I)Organization of work and property	I.1) Collective work model from the Jesuit reductions.	Bruxel (1978); Flores (1986); Flôres (2012); Freitas (1998); Langer (2005); Maurer (2010); Oliveira (2011a); Ornellas (1976); RIHGEB (1863); and Venturini (2009) (10 citations)
	I.2) Individualism: Spanish, Portuguese, Brazilian, Missionary and Sao-Borjense.	César (2002); Flores (1986); Flôres (2012); Langer (2005); Laytano (1983); Porto (2007); Putnam (2002); RIHGEB (1863); Rillo (1982); Rosa (2011); Saint Hilaire (2002); Silveira (1979); Torronteguy (1994); Vergara (2010) and Venturini (2009) (15 citations)
II)Productive activity	II.1) Fruit-growing: During the Jesuit introduction. And later as the "land of orange groves", with the nickname "papa laranja" (orange pope), until the decay of orange groves (citrus canker); and peach trees.	Baguet (1997); Cotrinoticias (1983); Flôres (2012); Silveira (1979); Saint Hilaire (2002); Dreys (1990); Gay (1980); Rodrigues (1982); Rosa (2011) (9 citations)
	II.2) Cattle farming: From the introduction of cattle on the east bank of the Uruguay River in 1635 by Jesuit priests Pedro Romero and Cristobal de Mendoza to the 21st century.	Bruxel (1978); César (2002); Dreys (1990); Flores (2004); Flôres (2012); Langer (2005); Maurer (2010); Marques (1987); Marques (2010); Rodrigues (1982); Rillo (1982); and Saint Hilaire (2002) (12 citations)
	II.3) Agriculture: From the Prussian-German	The Tribune (1969); Cotrinoticias (1983);

	immigrants at the end of the 19th century	Flôres (2012); Folha de Sao Borja
Categories	**Sub-categories**	**Sources of citations**
	with wheat, to the "production capital" of wheat, flax and linseed; to the current predominance of rice and soybeans.	(1981); Rodrigues (1982); Rillo (1982) (6 citations)
	II.4) Home industry, manioc flour and wheat flour production in the 19th century.	Flores (2004); Flôres (2012); Rodrigues (1982); Rillo (1982); Silveira (1979); (5 citations)
III) Income	III.1) Lack of income/lack of commercialization: poverty and misery from the "dispossessed" groups of the past (Indians, Africans, peasants, gaudèrios) of the 18th century to the rural "dispossessed" of today.	Abramovay (1998); Bakos (1982) Cardoso; Falleto (1979); Cèsar (2002) Cunha (2006); Flores (1986); Flôres (2012); Langer (2005); Laytano (1983);Machado (1999); Ornellas (1976);Pesavento (1988);Porto (2007); Rodrigues (1982);Rosa (2011); Silveira (1979); Torronteguy (1994); Venturini (2009); Waquil and Mattos (2002); Wilkison (2008); and Zarth (2006); (23 citations)
IV)Education	IV.1) Education in São Borja: from the illiterate of the 18th century, to schooling from 1880 onwards in the city and the districts of the campaign. Up to those with little schooling in the 21st century.	CSCJ (1992); Flôres (2012); Langer (2005); Silveira (1979); Torronteguy (1994); Venturini (2009); and Zero Hora (Oct.2012) (7 quotes)
V)Assistance Technical	V.1) Technical assistance in Sâo Borja: from the "experimental seed station" run by Franco Baglioni, through the intercession of Getúlio Vargas, to advice from cooperatives. After their decline, the cooperatives were assisted by the government, liberal professionals and companies.	A Tribuna (1969); Folha de Sâo Borja (1981); Cotrinoticias (1983); Rillo ' (1982); and Rodrigues (1982) (6 citations)
IV) Political leadership	IV.1) Caudillo culture and Colonelism. From the caudillos who invaded San Borja: Andresito Artigas in 1816 and Frutuoso Rivera in 1828. Right up to the continuity of caudillista and colonelist politics in the 20th century in the municipality.	Cardoso and Falleto (1979), Cèsar (2002), Flores (2004), Flôres (2012), Colussi (2008), Khanna (2009), Laytano (1983), Machado (1999), Marques (1987), O'donnell (1985), Pesavento (1988), Rillo (1982), Silveira (1979), Teixeira (2011), Torronteguy (1994), Venturini (2009) (16 citations).

Source: Prepared by the author based on Bardin (2011); Meireles and Cendón (2010); and Section 1.2.

Table 10 uses content analysis and citation analysis in the subcategories by authors and newspaper reports. These are where we found the topics most frequently covered in the historical survey:

- **1° Lack of income/lack of commercialization** - 23 citations. This is the realization that Sao Borja, after Iberian domination (Spanish and Portuguese), has never been and is not egalitarian in terms of income distribution. Especially in the relationship between large properties and small ones, and historically small properties have been relegated only to the role of aggregates, with little political/public attention. The incentives are ad hoc, just to help "the dispossessed" with palliative measures. They don't tackle the causes of poverty, there is no political interest;

- **2° Caudillesque culture, coronelismo** - 16 citations. The second reinforces the individualistic profile of those in power (subjects - men, white, with possessions, weapons and power x objects - the dispossessed: indigenous, Afro-descendants, missionaries, gaudérios, peasants, soldiers, illiterate, poor, family farmers, little education). They are populists, with actions to preserve themselves in power (for their own interests and self-promotion), increasing their possessions and with political influence (cacicado, clientelism-familyism). They don't aim to do good for the dispossessed, who are historically a mass of maneuvering. They use the politics of patronage, the value of prestige and the culture of silence (violence);

- **3rd Individualist Model** - 15 citations. The individualist model is the root of the municipality's income problems and historically low development. A few dominate and their interests prevail to the detriment of the vast majority of the population. The research into socio-productive profiles details in detail and in depth the relationships between the individualist model and the lack of income, poor education, lack of attitude; in other words, low development on the properties analyzed, both qualitatively and quantitatively in the municipality. The research into the analysis of human and social capacity also explains the factors that have a negative impact, such as: young people leaving for other cities, i.e. low confidence, little entrepreneurship, low cooperativism, low schooling and political leaders who are caudillos, coronelistas; all of which are directly influenced by the individualist culture;

- **4th Cattle farming** - 12 citations. Historically, cattle farming was the most important economic activity in the municipality from the beginning of the 17th century until the middle of the 20th century. After the middle of the 20th century, agriculture prevailed in terms of economic importance. For the historical factor it is well cited;

- **5th Collectivist Model** - 10 citations. Being the beginning of the history of San(Sao) Borja and for the fascination that this model impressed on the authors, and the vast literature on the Seven Peoples of the Missions - Missionary period. Today, according to this work, collectivist foundations would be vital for family farming to develop. The lack of a collectivist model is seen as limiting the development of research into socio-productive profiles and the analysis of human and social capacity;

- **6° Fruticultura** - 9 citations. Fruit growing historically began with the Jesuits in the missionary period and was widely used, not economically, until the end of the 19th century. In the 20th century, citrus canker was eradicated in 1983. It is highlighted in the research into socio-productive profiles as a suggestion for revitalizing productive activity for family farmers, as it is a historical alternative that fosters income, requires the unity of producers and uses a lot of labor (family and non-family). However, there are risks that can affect the activity related to aerial herbicides. It requires a scale of production (volume), technical assistance with rural extension, fruit quality and the search for commercialization in various buyer markets (clients) - collective entrepreneurship;

- **7° Education** - 7 quotes. Education started late for the general population, only in the 19th century. Today the municipality is considered average by public and independent educational assessment bodies. The research into socio-productive profiles goes into great detail about the low level of education of the head of the family on the properties analyzed, and education is considered to be a challenge in the municipality. In the human and social capacity analysis research, low schooling is considered a limitation;
- **8th Agriculture** - 6 citations. With the arrival of European immigrants at the end of the 19th century to work in agriculture, and then at the beginning of the 20th century and in the middle of it, these immigrants and their descendants changed the productive vocational profile of Sao Borja. They whitened the population and, from the middle of the 20th century, agriculture took over as the municipality's economic leader. It is rarely mentioned due to the lack of authors on Sao Borja from the 1990s onwards. However, it is of vital social, economic and political importance;
- **8th Technical Assistance** - 6 citations. Considered recent in the history Brazil, Rio Grande do Sul and Sâo Borja. But it is the fundamental support for family farming. In the survey of socio-productive profiles, it was considered to be below the capacity to provide services and the needs of the properties - owners of family and non-family farms up to 100 ha in Sâo Borja (mainly the public one, because the private one is paid for and few properties have the financial means to pay for it). This is a challenge for family farmers. A small number of references were found, so there are few citations;
- **9° Cottage industry, making manioc flour and wheat flour** - 5 citations. Basically historical references. Today, cottage industries are considered non-existent. There are no manioc flour factories. There is a wheat mill, but it serves the local market.

3.3 THE SOCIO-PRODUCTIVE SITUATION OF SMALL AND MEDIUM-SIZED rural properties in the municipality of sÂO bórja - rS

In this section we present the main results and comments from the field research with 248 rural properties of up to 100 hectares in the municipality of Sâo Borja - RS.

In accordance with the methodological procedures defined in Section 2.3, the form for the survey of the socio-productive profile of small and medium-sized farms was drawn up and applied. The complete tabulation of the data collected can be found in Appendix A on pages 138-144.

In order to establish a general characterization of the situation surveyed, a summary of the main results is presented in Table 11.

Table 11 - Main results of the survey of the socio-productive profile of small and medium-sized rural properties in the municipality of Sâo Borja-RS

Features	Proportion of occurrence in relation to properties sampled (%)
Income supplement:	
- Retirement and sickness benefits	64,13
Third-party labor	23,2
Youth migration	57,3
Family farming	Yes: 85.1
	They have DAP: 34.3
Electricity	Yes: 99.6
Water:	
- Human consumption	Good: 63.7
- For creation	Scarce: 41.5
- For tillage	Non-existent: 61.3
- Community	Yes: 67.7
Family income	Up to R$1,194.53: 67.3
Level of education of the head of	
family:	
- Illiterate and up to 3rd grade	28,6
- Primary 4ª	33,9
Location in own area	71,9
Main agricultural crops	Soybeans, wheat, corn (grain, cob, livestock), rice and sorghum
Agricultural technical assistance	Yes: 11.3
Use of inputs	Low: 24.2 and regular: 13.3
Investments to increase	
productivity:	
- Volume of resources required	High: 71.9
- Suitable conditions for investment	Low: 84.2
Satisfaction with marketing	Yes: 3.2
farming	
Use of machinery:	
- In partnership	
- Own	11,7 19
- Community agricultural patrol	27,8
Agricultural crop costs	Own resources: 48.4
Subsistence crops (manioc and secondarily strawberry/pumpkin, watermelon, melon and sweet potato):	Yes: 7.7
- Technical assistance	Yes: 26.2
- Marketing production	Yes: 15.7

- Satisfaction with marketing	
Vegetables:	
- Produce	Yes: 83.1
- Technology	Low: 65.3
- Composting	Yes: 23.0
- Organic fertilizer	Yes: 84.7
- They sell their produce	Yes: 21.4
- Satisfaction with marketing	Yes: 12.9
Cattle:	
Features	**Proportion of occurrence in relation to properties sampled (%)**
- Own consumption	Yes: 41.5
- Sales ratio/Total number of animals	24.4 (Low)
- Satisfaction with marketing	Yes: 14.9
Tambo de leite:	Yes: 5.2
- Periodic technical assistance from a professional	Yes: 6.1
- Emater	Yes: 10.5
- Coptec	Yes: 13.3
- Satisfaction with marketing	Own 12,5
- Use of machinery	In partnership: 1.2
	None: 20.2
Small animals:	
- Sheep	Own consumption: 29.8
Sales ratio/Total number of animals	18.3 (Low)
- Pigs (largest income generator)	Own consumption: 48.8
- Satisfaction with marketing	Yes: 12.9
- Costs	Own resources: 49.5
- Technical assistance Periodic visits by a professional	Yes: 3.6
- Emater	Yes: 2.8
- Coptec	Yes: 2.8
Birds:	
- Hens and roosters	Own consumption: 83.9
- Satisfaction with marketing	Yes: 17.3
Aquaculture:	
- Weirs and ponds	Yes: 9.3
- Main species	Carp: 33.3
- No. of producers marketing; and No. of those satisfied with marketing	3; e 1
Beekeeping:	
- Boxes	N°: 505
- Satisfaction with marketing	Yes: 10.1
Orchard - Fruit growing:	Orange tree: 21,8;
- Species	bergamot tree: 20,6;

- Technical assistance	peach tree: 12,3; e
- City Hall	hose: 5
- Emater	Yes: 1.9
- Coptec	Yes: 6.9
- Satisfaction with a	Yes: 6.3
marketing	Yes: 13.3
Member of a cooperative or union	Yes: 34.7
of producers:	Collective sales: 33.9; collective purchases: 28.6; sales to
Agree with:	
Features	**Proportion of occurrence in relation to properties sampled (%)**
	Cooperatives: 29.4; and
	Sales to associations:
	24,6
Main needs for	
increased competitiveness:	
- Irrigation	Yes: 39.1
- Technical assistance	Yes: 33.5
- Financial resources	Yes: 31.8
- Weirs	Yes: 29.4
- Soil correction	Yes: 23.8
- Artesian wells	Yes: 14.1
- Marketing	Yes: 13.7
Main threats à	
competitiveness:	
- Drought, drought	Yes: 73.0
- Lack of income	Yes: 48.4
- Aerial herbicide	Yes: 39.1
- Lack of marketing	Yes: 19.3
- Abigeato	Yes: 12.9
- Indebtedness	Yes: 10.5
Main short courses	
interest:	
- Irrigation	Yes: 24.2
- Bovine farming	Yes: 23.8
- Rural management	Yes: 14.9
- Food industrialization and	Yes: 13.7
preserves	
- Making the most of food	Yes: 12.9
- waste	
- Horticulture	Yes: 11.7
- Rural entrepreneurship	Yes: 11.3

- Fruit growing	Yes: 10.1
- Poultry farming	Yes: 10.1
Product marketing	
homemade:	
- Milk	Yes: 12.9
- Eggs	Yes: 10.0
- Cheese	Yes: 9.7
- Sweets	Yes: 9.7
- Cuca	Yes: 7.0
- Rapadura	Yes: 5.6

Source: Author's elaboration.

Table 11 shows that of the 248 properties sampled, 71.9% occupy their own land, 85.1% are predominantly family farms and 14.9% are made up of a small number of corporate farmers. However, with regard to the declaration of suitability for the National Program to Strengthen Family Farming (PRONAF), it can be seen that only 34.3% have the documentation to be able to access bank credit.

The economic livelihood of the properties is precarious. Family income for 67.3% of the families is up to R$1,194.53. A significant portion of this income comes from pensions (46%) and sickness benefits (18.13%), earned by 46 and 18.13% of the people living there, respectively. Even so, 23.2% of the properties make family labor available to third parties. This indicates a shortage of work opportunities and solidary, collective actions that require relationships of reciprocity. And at the root of this context is the low level of education of the heads of household, 62.5% of whom have up to 4th grade education, including a good number of illiterate people.

The main agricultural crops grown are soybeans, wheat, corn (grain, cobs, livestock), rice and sorghum. This production is paid for with their own resources and only 11.6% of the properties receive funds from companies and banks.

As for the availability of some factors of production, the situation is worrying in the case of water for livestock consumption, which is scarce on 41.5% of the properties, and for use on crops, which is non-existent on 61.3%, seriously compromising the development of these activities. With regard to water for human consumption, community water and electricity, the situation is satisfactory.

The use of inputs is unsatisfactory in 37.5% of the farms and only 42% of them use machinery in production. In 19% of the farms, the machinery is owned and in the others it is used in partnership by 11.7% or rented by 11.3%. Even so, only 27.8% of families use agricultural patrol machinery, revealing an insufficient degree of technification.

Technical assistance is not provided, i.e. it is non-existent, in 78.1% of the production units.

In the face of this less than encouraging picture, there are still a number of other worrying effects. For example, there is widespread dissatisfaction on farms that sell agricultural products, according to 96.8% of them. Another effect is the intense flight of young people from the countryside (rural exodus), which 57.3% of families reported.

In order to increase productivity, the level of investment required is considered high by 71.9% of respondents. On the other hand, the conditions under which such investments can be made are considered very inadequate by 84.2% of the properties.

Subsistence crops include manioc and, secondarily, strawberry/pumpkin, watermelon, melon and sweet potato. Vegetables are also grown with organic fertilizer on 83.1% of the properties. In 41.5% of the units, cattle are raised for their own consumption, and only 24.4% of the total herd is sold. In the case of sheep, these ratios are 29.8% and 18.3% respectively. In the case of small animals as a whole, in addition to being food for 29.8% of families, pigs are the biggest source of income on farms with small animals. The contribution of poultry to the diet is also very important for almost 84% of families.

For their part, aquaculture, beekeeping and fruit-growing are of little importance in the production system of family farming activities. Sao Borja was once called the "land of the orange groves" and the people of Sao Borja were known as the "orange pope". Today, this activity is in sharp decline due to competition from rice, soya/wheat and the use of aerial herbicides.

Generally speaking, in subsistence farming (manioc and, secondarily, strawberry/pumpkin, watermelon, melon and sweet potato), vegetable and fruit growing, cattle breeding (and milk production), sheep, pigs, poultry and bees - the proportion of properties satisfied with marketing is very small, ranging from 10.1 to 17.3%. Dairy farming is poorly served due to the decrease in production caused by the drought, which reduced supply below the minimum quota required by the buying company, which closed its business in the region. In the case of aquaculture, although 23 properties have dams and ponds, only 3 of them market their production and, of these, only one says it is satisfied with the sale of the fish. With regard to the marketing of homemade products, this is practiced by 13% of the units (milk); around 10% (eggs, cheese and sweets); and around 7% (cuca and rapadura).

To make matters worse, only 34.7% of farms are members of cooperatives or producer unions, but with very different rates of agreement regarding marketing and collective purchases and sales to cooperatives and/or associations.

Using lexical analysis processed in the SPHINX *software,* the perception of the main needs to increase competitiveness was computed. Farmers pointed to irrigation (39.11%), technical assistance (33.46%), financial resources (31.85%) and dams (29.43%). Soil correction (23.8%), artesian wells (14.1%) and marketing (13.7%) were also mentioned.

The priority concern is the survival of crops and livestock, given the frequent and prolonged recurrence of droughts and dry spells. It also confirms the need for solutions to the lack of technical assistance and the contribution of financial resources for investments in soil correction and damming.

With regard to the main threats to competitiveness, the lexical analysis reveals that 73% of the units point to drought/drought, 48.4% to lack of income and 39.1% to aerial herbicides; as well as, but with much less emphasis, lack of marketing, land grabbing and indebtedness. During the field surveys, it became apparent that, due to the drought/drought and the consequent depletion of sources of income, properties are being abandoned and uninhabited houses turned into "taperas". Another very worrying threat to health is aerial herbicides, the launch of which was recorded by the researcher. After the herbicide was applied, the plane flew over the houses in the Caçaca settlement.

Finally, when it came to indicating the main short-term courses of interest, farmers chose, in the range of 24.2 to 13%, in descending order, training in irrigation, cattle farming, rural administration, food processing/conserving and food waste; and, between 10 and 12%, horticulture, rural entrepreneurship, fruit growing and poultry farming.

In order to obtain more specialized and in-depth results and findings, some cross-references of interest to the research were processed in the SPHINX *software.*

Firstly, the situations "are you a member or a cooperative member?", "collective marketing", "collective purchases", "sales to cooperatives" and "sales by associations" were compared (Table 12). The opinions "indifferent", "disagree", "totally disagree" and "no" were given simultaneously by 68.8% of the interviewees. This attitude is further strong evidence of the individualistic model, which was pointed out in 15 sources in the content and citation analysis carried out in Section 3.2 (Table 10, item I.2).

Table 12 - Multiple analysis: are you a member or a cooperative member? ; collective marketing ; collective purchases ; sales to cooperatives and sales by associations

Results			
[Sale by Associations]	Category 3	Indifferent 171	14,4%
[Are you a member or a cooperative member?]	Category 2	No 152	12,8%
[Collective shopping]	Category 3	Indifferent 151	12,7%
[Cooperative Sales]	Category 3	Indifferent 146	12,3%
[Collective marketing]	Category 3	Indifferent 135	11,3%
[Are you a member or a cooperative member?]	Category 1	Yes 86	7,2%
[Collective marketing]	Category 4	I agree 76	6,4%
[Cooperative Sales]	Category 4	I agree 66	5,5%
[Collective shopping]	Category 4	I agree 64	5,4%
[Sale by Associations]	Category 4	I agree 57	4,8%
[Cooperative Sales]	Category 2	I disagree 15	1,3%
[Collective marketing]	Category 2	I disagree 11	0,9%
[Collective shopping]	Category 1	Strongly Disagree 9	0,8%
[Collective marketing]	Category 5	Totally agree 8	0,7%
[Collective shopping]	Category 2	I disagree 8	0,7%
[Collective marketing]	Category 1	Strongly Disagree 7	0,6%
[Collective shopping]	Category 5	Strongly agree 7	0,6%
[Cooperative Sales]	Category 5	Strongly agree 7	0,6%
[Cooperative Sales]	Category 1	Strongly Disagree 5	0,4%
[Sale by Associations]	Category 2	I disagree 4	0,3%
[Sale by Associations]	Category 5	I agree Fully 4	0,3%
[Sale by Associations]	Category 1	Strongly Disagree 3	0,3%
	TOTAL	1192	100%

Source: Author's elaboration.

Cross-referencing the estimated monthly income, the level of investment requirement and the condition of investment in the property (Table 13) shows that 167 properties, 67.34% of which fall into income brackets between R$276.70 and R$1,194.53 - categories N°1"E", N°2"D", N°3"C2" and N°4"C1". This means that these units do not earn two national minimum wages, considered by IPEA (2003) to be the minimum necessary to survive with dignity. This finding by IPEA refers to urban populations. In rural areas, the situation is worse, since they need to invest, pay for inputs, animals and machinery, in addition to meeting the other expenses of the property - Waquil; Mattos (2002) classify this situation as "rural poverty". Under these conditions, only 81 properties (32.66%) are able to make any investments, cover costs and expenses and earn a surplus.

Also in this analysis, it can be seen that in the units with an estimated monthly income of up to R$1,194.53, 55 of them consider the level of demand for their investments to be high and 62 consider their ability to invest with their own resources to be insufficient.

Table 13 - Cross-references between the estimated monthly income, the level of investment requirement and the condition of investment in the property

Estimated Income	Investment Exgency Level	Condition of Investment in Property
Category n° 1,"E>R$276.70" (17)	Category n° 3, High (3)	Category n° 1, Bass (3)
Category n° 2,"D>R$484.97" (33)	Category n° 3, High (13) Category n° 2, Medium (2)	Category n° 1, Bass (15)
Category n° 3,"C2>R$72626" (59)	Category n° 3, High (19) Category n° 2, Medium (5)	Category no. 1, Bass (23) Category n° 2, Medium (1)
Category n° 4,"C1>R$1194.53" (58)	Category n° 3, High (20) Category n° 2, Medium (3)	Category n° 1, Bass (21) Category n° 2, Medium (2)
Category n° 5,"B2>R$2012,67" (45)	Category n° 3, High (13) Category n° 2, Medium (7)	Category n° 1, Bass (18) Category n° 2, Medium (2)
Category n° 6,"B1>R$3479.36" (21)	Category n° 3, High (10) Category n° 2, Medium (6)	Category n° 1, Bass (11) Category n° 2, Medium (5)
Category n° 7,"A2>R$6563.73" (7)	Category n° 2, Medium (4) Category n° 3, High (2)	Category n° 1, Bass (3) Category n° 2, Medium (3)
Category n° 8,"A1>R$9733.47" (8)	Category n° 2, Medium (4) Category n° 3, High (2) Category n° 1, Bass (1)	Category n° 2, Medium (4) Category n° 1, Bass (2) Category n° 3, High (1)
SET (248)	**Category n° 3, High (82)** **Category n° 2, Medium (31)** **Category n° 1, Bass (1)**	**Category No. 1, Bass (96)** **Category n° 2, Medium (17)** **Category n° 3, High (1)**

Source: Author's elaboration.

In order to ascertain the association between the level of education of the head of household and the various types of technical assistance received by the property, we considered the assistance given to agricultural crops and subsistence crops, the advice given to large animals by Emater and Coptec, the provision of occasional services for large animals, periodic visits by professionals and the advice given by Emater and Coptec technicians for small animals (Table 14).

Table 14 - Multiple analysis: level of education of the head of household and the various types of technical assistance received by the property

Level of education of head household	Technical assistance for agricultural crops	Technical assistance for subsistence farming	Emater advises on large animals	Coptec advises on large animals	One-off services for large animals	Professionals regularly visit small animals	Emater advises small animals	Coptec advises small animals
Category n°1 Illiterate up to 3rd grade (71 owners)	Yes (2) No (32)	Yes (2) No (62)	Yes (2) No (18)	Yes (11) No (3)	Yes (5) No(27)	Yes (4) No (19)	Yes (3) No (11)	Yes (3) No (5)
Category n°2 4ª Incomplete primary school (84 owners)	Yes (7) No (34)	Yes (5) No (75)	Yes (5) No (18)	Yes (11) No (4)	Yes (12) No (26)	Yes (1) No (31)	Yes (3) No (21)	Yes (3) No (6)
Category n°3 Complete primary education (60 owners)	Yes (12) No (23)	Yes (6) No (44)	Yes (4) No (8)	Yes (3) No (5)	Yes (6) No (14)	Yes (2) No (13)	No (10)	Yes (1) No (4)
Category n°4 Secondary Education (22 owners)	Yes (4) No (9)	Yes (3) No (15)	Yes (1) No (3)	Yes (1) No (0)	Yes (3) No (3)	Yes (2) No (6)	No (7)	N/C
Category n°5 Complete university degree (11 owners)	Yes (3) No (2)	Yes (3) No (2)	Yes (3) No (1)	N/C	Yes (3) No (1)	No (2)	Yes (1) No (1)	N/C
Set 248 Properties	Yes (28) No (100)	Yes (19) No (201)	Yes (15) No (48)	Yes (26) No (12)	Yes (29) No (71)	Yes (9) No (71)	Yes (7) No (50)	Yes (7) No (15)

Source: Author's elaboration.

Table 14 shows the low availability of technical assistance and/or low interest in accessing it. A very small number of owners obtain some kind of technical assistance. In the first three categories, this occurs in the large animal advisory services provided by Coptec. In the case of farmers with high school and college degrees, the lack of assistance is widespread and more intriguing as to why they are left on the sidelines, since these are farmers with a higher level of education.

Under these conditions, there is a blatant lack of effective technical assistance, both public and private, for all 248 properties surveyed, regardless of the level of education of the head of household. Undoubtedly, rural technical assistance is a major challenge facing rural actors in the municipality of Sao Borja.

In order to examine the education levels of farmers in the different income strata, these variables were crossed (Table 15).

Table 15 - Multiple analysis: estimated monthly income and level of education of the head of household

Estimated Income	Level of Education Head of Household
Category n° 1,"E>R$276.70" (17)	Category No. 2, 4ª Elementary School + (8 ; 47.1% ; 1.39) Category No. 4, Complete High School + (2 ; 11.8% ; 1.33) Category n° 3, Fundamental Compl eto + (4 ; 23.5% ; 0.97) Category 1, Illiterate up to 3rd grade + (3 ; 17.7% ; 0.62)
Category n° 2,"D>R$484.97" (33)	Category No. 1, Illiterate up to 3rd Grade + (12 ; 36.4% ; 1.27) Category No. 3, Complete Elementary + (10 ; 30.3% ; 1.25) Category No. 2, 4th Grade + (10 ; 30.3% ; 0.90) Category No. 4, Complete High School + (1 ; 3.0% ; 0.34)
Category n° 3,"C2>R$726.26" (59)	Category No. 1, Illiterate up to 3rd Grade + (20 ; 33.9% ; 1.18) Category No. 3, Complete Elementary + (16 ; 27.1% ; 1.12) Category No. 2, 4th Grade + (19 ; 32.2% ; 0.95) Category No. 4, Complete High School + (3 ; 5.1% ; 0.57) Category No. 5, Complete Higher Education + (1 ; 1.7% ; 0.38)
Category n° 4,"C1>R$1194.53" (58)	Category No. 5, Complete Higher Education + (4 ; 6.9% ; 1.56) Category No. 1, Illiterate up to 3rd Grade + (18 ; 31.0% ; 1.08) Category No. 2, 4th Grade + (20 ; 34.5% ; 1.02) Category No. 3, Complete Elementary + (12 ; 20.7% ; 0.86) Category No. 4, Complete High School + (4 ; 6.9% ; 0.78)
Category n° 5,"B2>R$2012,67" (45)	Category No. 2, 4th Grade + (22 ; 48.9% ; 1.44) Category No. 4, Complete High School + (4 ; 8.9% ; 1.00) Category No. 5, Complete Higher Education + (2 ; 4.4% ; 1.00) Category No. 1, Illiterate up to 3rd Grade and Elementary + (12 ; 26.7% ; 0.93) Category n° 3, Fundamental Compl eto + (5 ; 11.1% ; 0.46)
Category n° 6,"B1>R$3479.36" (21)	Category No. 4, Complete High School + (3 ; 14.3% ; 1.61) Category No. 3, Complete Elementary School + (7 ; 33.3% ; 1.38) Category No. 5, Complete Higher Education + (1 ; 4.8% ; 1.07) Category No. 1, Illiterate up to 3rd Elementary School + (6 ; 28.6% ; 1.00) Category No. 2, 4th Elementary School + (4 ; 19.1% ; 0.56)
Category n° 7,"A2>R$6563.73" (7)	Category No. 4, Complete High School + (3 ; 42.9% ; 4.83) Category 5, Superior Compl eto + (1 ; 14.3% ; 3.22) Category No. 3, Fundamental Compl eto + (3 ; 42.9% ; 1.77)
Category n° 8,"A1>R$9733.47" (8)	Category n° 5, Superior Compl eto + (2 ; 25.0% ; 5.64) Category No. 4, Complete High School + (2 ; 25.0% ; 2.82) Category n° 3, Fundamental Compl eto + (3 ; 37.5% ; 1.55) Category No. 2, 4th Grade + (1 ; 12.5% ; 0.37)
SET (248)	**Category No. 2, 4ª Elementary School (84)** **Category No. 1, Illiterate up to 3rd grade (71)** **Category No. 3, Elementary C omplete (60)** **Category No. 4, Complete High School (22)** **Category No. 5, Higher Education Completed (11)**

Source: Author's elaboration

Table 15 (last row, second column) shows that the predominant levels of education are 4th grade (84 owners) and illiterates up to 3rd grade (71 owners), making up 62.5% of all heads of household with an incipient level of education.

The first column of Table 15 also shows that the number of farmers went from 17 to 59, from category

1 to 3. In the next two categories, the number of farmers remained at levels not very different from those in category 3. In the last 3 strata, with the highest incomes, the number of owners fell from 21 to 8. In order to investigate the evolution recorded in the first three income categories, Table 16 was drawn up, linking these strata with the categories representing the different levels of education of the heads of household.

Table 16 - Multiple analysis: education level of the head of household in the first 3 strata of estimated monthly income

Instruction Categories	**Category n° 1,"E>R$276.70" (17)**	**Category n° 2,"E>R$484.97" (33)**	**Category n° 3,"E>R$726.26" (59)**	**Total**
Category 1 Illiterate up to 3rd grade (71 owners)	3	12	20	35
Category 2 4ª Incomplete primary school (84 owners)	8	10	19	37
Category 3 Complete primary education (60 owners)	4	10	16	30
Category No. 4 High School Completed (22 owners)	2	1	3	6
Category 5 Complete university degree (11 owners)	-	-	1	1
Total	17	33	59	109

Source: Author's elaboration.

The same table shows that the total number of landlords increased by 247% (from 17 to 59) from the first to the third income stratum, while remaining relatively stable from the first to the third category, referring to level of education. This performance reflects a good influence of education level on family income, albeit within a limited range. Even so, this is a stimulating potential response for the planning and implementation of actions to expand and improve formal education, especially professional education.

This confirms Figueiredo's (2012) view of the persistence of huge obstacles due to the lack of training for rural producers in Rio Grande do Sul and Sao Borja. For Tietbohl (2012), it is necessary to overcome the accumulated deficiency in formal education, with training that the farmer can use in rural activity and overcome the educational backwardness in relation to the evolution of technology.

In order to analyze some of the aspects inherent in the units run by heads of household with a university degree, we selected elements of interest from this stratum, especially with regard to the relationship between individualism and the non-use or partial use of technical assistance (Table 17).

Table 17 - Some aspects inherent to units run by heads of household with a university degree and identification of their characteristics in terms of management profile

Estimated Income	Level of education of the head of	Property description	Comments on the use of technical

	household (occurrences)		assistance and profile
R$ 726.26 (59 properties)	1 complete university degree	- Property no. 152 - 22 hectares. Rural producer. Activities: agriculture, vegetable gardening, sheep farming, poultry farming and the main beef and dairy farming. Few cattle sales. Loss of corn crop due to drought. No soil analysis or production costs, low use of inputs, no machinery. Lost subsistence crop production. They don't manage their property properly. She is a member, but her answers to the questions collective marketing, collective purchases, sales to cooperatives and sales by associations are always indifferent.	- Survey respondent 152 - Diversified producer, but low production and marketing. Without technical assistance, admits lack of income from threats. Traditional cattle farmer. **Individualist profile**.
R$ 1,194.53 (58 properties)	4 complete university degree	• Property no. 128 - 30 hectares. Activities: cattle farming, sheep farming and subsistence crops. Rural producer and has another professional activity. Area in condominium.1 Retired member of the family. Has some degree of management. Not a member or cooperative member, but agrees with collective marketing, collective purchases and sales by associations. Indifferent about selling to cooperatives. Property is being structured. • Property no. 197 - 60 hectares. Activities: subsistence crops, horticulture, sheep farming and cattle farming. He is a retired farmer with the surname of the first Portuguese families to receive land in the 19th century in Sâo Borja. He lost his subsistence crops with the	• Respondent 128 - No technical assistance. Traditional farmer. Lacks attitude. Slow investments because they don't fit into Pronaf. Not entrepreneurial. **Individualistic profile**. • Respondent 197 - No technical assistance. Traditional cattle farmer. Lacks attitude due to retirement. Admits lack of income from threats and has lost animals due to lack of pasture. **Individualist profile**.
		dry. Cattle production exclusively in native fields. Few cattle sales and marketing with intermediaries. Has minimal management skills. They are not members or cooperatives and have always given indifferent answers to the questions collective marketing, collective purchases, sales to cooperatives and sales by associations. • Property no. 250 - 6 hectares. 1 retired person. Activities: agricultural crops, subsistence crops, vegetables, cattle, pig and poultry farming. Rural producer of German origin, only sells pigs, eggs and homemade	• Respondent 250 - They receive technical assistance from Emater on corn every six months, and small animals. Lack of income is a threat. They have a cooperative profile but have not marketed their cattle production. It is diversified. Lack of production for marketing. Therefore, **an individualist profile**. • Respondent No. 253 - No technical advice because there is no focus on production and marketing. Threat is

		products. He grazes annual winter pasture and perennial summer pasture. He is a cooperative member (Cotrisal) and agrees to collective marketing, collective purchases, sales to cooperatives and sales by associations. • Property no. 253 - 18 hectares. Former rural producer. 2 pensions, husband with health problems. They were robbed on the rural property. Grows subsistence crops, vegetables, fish; all without a focus on production and marketing.	insecurity of family property. **Individualistic profile.**
R$ 2,012.67 (45 properties)	2 complete university degree	Property No. 3 - 5 hectares. Rural producer. Activities: subsistence crops, vegetables and fruit growing. 2 workers during the year. He markets all his activities: direct to consumers, restaurants and fruit shops. It is associated with market traders. They have management. Agrees with collective marketing, collective purchases and sales by associations. Is indifferent to selling to cooperatives. - Property No. 4 - Rural producer. 18.5 hectares. 8.5 hectares owned and 10 hectares rented. Dairy farming. They have annual winter and summer pasture, as well as perennial pasture. Supplements the animals with feed and hay. He funds the activity with his own resources and Banco do Brasil. He is a member of Nhupoleite. Does management. They agree with collective marketing, collective purchasing, fully selling to cooperatives and agree with selling	• Respondent No. 3 - Is an individual entrepreneur and a collective entrepreneur. They have technical assistance from Emater. Needs infrastructure and the threat of aerial herbicides for their activities. **Non-individualistic profile.** • Respondent 4 - He is an individual entrepreneur. They have a veterinary doctor in the village who visits periodically and assistance from Emater. They are specialized, but they don't have a milking machine. They are a little dissatisfied with the remuneration for their milk. Needs: to reduce nutrition costs, agricultural patrol and sugar. The threats are the climate, aerial herbicides and cattle rustling **Non-individualistic profile.**
		by the associations.	
R$ 3,479.36 (21 properties)	1 complete university degree	- Property no. 22 - 60.5 hectares. 55.5 hectares owned and 5 hectares leased. Rural producer with a traditional family name in Sao Borja, he is an agronomist. Has other commercial activities. 2 workers with a formal contract. Activities: agricultural crops, subsistence crops, vegetables, dairy cattle, poultry, beekeeping. All the activities carried out are sold directly to producers, consumers, the public market and through associations. A member of Nhupoleite. Supplements the herd with feed, sorghum silage and pasture hay. They pay for the stall with their own funds and	- Respondent no. 22 - He is an individual entrepreneur, a collective entrepreneur, an associate and a cooperative member. He is an agronomist and has veterinary technical assistance. The property is diversified. Needs a dam or reservoir, irrigation equipment and skilled labor. Threats Considers the roads to be poorly maintained, criticizes the marketing of milk and the criteria for demanding milk quality without technical assistance from the buying company. **Non-**

		funds from the banks of Brazil and Sicredi. They have their own advanced milking machine. They have management. They agree to collective marketing, collective purchases, sales to cooperatives and sales by associations.	**individualistic profile.**
R$ 6,563.73 (7 properties)	1 complete university degree	- Property no. 91 - 16 hectares. Tenant farmer. Has another main activity. Focus on agricultural crops: corn, sorghum and wheat. Machinery in partnership. Costs activities with own resources and uses resources from Banrisul and Sicredi banks. Not a member or cooperative member, but agrees with collective marketing, collective purchases, sales to cooperatives and sales by associations.	- Owner no. 91 - A rural entrepreneur. Specialized individual entrepreneur. Has a hired agronomist on a regular basis. Needs soil correction, irrigation and financial resources. The threats to their property include: drought, agricultural prices and lack of labor. **Non-individualistic profile.**
R$ +9.733,47 (8 properties)	2 complete university degree	• Property no. 184 - 11 hectares. Rural producer, from a family of Italian origin, he is an agronomist. 1 pension. 3 formal workers and 14 casual workers. Activities: subsistence farming, vegetables, sheep farming, poultry farming and the main activity is fruit growing. They have their own machinery. Very satisfied with viticulture. Are cooperative members of Agrocooper. They fully agree with collective marketing, collective purchasing, selling to cooperatives and agree with selling through associations. • Property no. 236 - 100 hectares. Rural producer with a traditional family name in Sao Borja. 2 pensions. 1 salaried worker and	• Owner no. 184 - He is a rural entrepreneur. He is an individual entrepreneur and a cooperative member. They have permanent technical assistance in viticulture from international technicians and consultants. Needs financial resources, machinery appropriate to the size of the property and sources of supply of seedlings. He lists the threats as climate problems and the cultural problem (individualism) of Sao Borja society. **Non-individualistic profile.** • Owner no. 236 - She is a rural entrepreneur. Individual entrepreneur. Her property is diversified and specializes in milk. They have
		1 casual worker per year. Activities: agricultural crops, vegetables, sheep farming and the main dairy farming. Extremely organized, planned and methodically cost-controlled property. Own machinery and funding from own resources and from Banco do Brasil. It uses rotational grazing and is a benchmark in irrigation systems. They have annual winter and summer pasture, as well as perennial pasture. He agrees with collective marketing and collective purchases. But he is indifferent to sales to cooperatives and sales by associations.	advice from a veterinarian in the field and from Emater. They need skilled labor, continuous, generalist technical assistance and better remuneration for agricultural products. Threats include drought and limited water for irrigation. **Non-individualistic profile.**

Total 248 properties	Total 11		

Source: Author's elaboration.

The first five owners listed in Table 17, in the individualist management profile (PI), are traditional cattle farmers and rural producers, of both sexes. In general, they don't manage their property, people and means of production properly. They don't use or only minimally use current agricultural and livestock techniques, such as electric fences, rotational grazing, mineralization (by different animal categories and ages) and invest timidly in machinery (no-till), improved pastures and animal supplementation. They make little or no use of inputs. They don't have technical assistance or receive it occasionally. They have no management and when they do make notes, they don't use them as input for decision-making and implementing actions. They are the cause, due to their lack of action, and the "victims" due to their lack of management of climate effects, and are therefore more susceptible to droughts and storms due to their lack of planning, organization and action. They interact minimally with their neighbors, the locality and society. They promote minimal production and marketing. They are resistant to innovation and averse to the risks of the activity. They invest minimally, have insufficient returns and, over the years, dilapidate their assets through immobility. They are skeptical and reactive. Their models are still from the 19th century. In short, they reflect the individualistic culture in the rural area, according to Vergara (2010), especially in terms of individualism and education.

For Capra (1982), who considers the holistic or systemic method to be the new paradigm, as opposed to the traditional scientific method, rural properties start to progress when they stop being managed in an individualistic way. This change, described by Capra as a "tipping point", begins to manifest itself on the farms investigated, starting with an estimated monthly income of R$2,012.67 (Table 17).

Among the 6 (six) owners classified in Table 17 as having a non-individualistic management profile (NIMP) are the individual entrepreneur, the collective entrepreneur and the active associationist and cooperativist, with satisfactory operational skills. They are distinguished by the correct management of the property (income, costs, expenses and investments); and appropriate management (health, animal and agricultural) of the factors of production and people management. They use current agricultural and livestock techniques, such as electric fences, rotational grazing and mineralization differentiated by animal category and age. They invest in infrastructure, machinery and no-till farming, use plant consortiums, crop rotation, improved pastures, animal supplementation and irrigation. They use inputs or organic production and make use of technical assistance. They establish partnerships. They organize themselves, draw up plans and implement them. They operate in a diversified and specialized way. They produce and market locally, regionally and globally. They interact with data and information, with the external environment, using computerization and connectivity. They have attitude, innovate, are proactive and take risks. Their actions converge to consolidate them as rural entrepreneurs. They represent a counterpoint to the individualistic culture in rural areas.

A similar procedure was extended to the other 237, applying content analysis (BARDIN, 2011) to the same managerial profiles analyzed in the case of heads of household with a university degree, plus another profile called "individualist+ intermediate" (PII). This profile covers units in the transition phase to the non-individualist profile. The intention is to capture the occurrence, at this stage, of management initiatives, access to technical assistance, increased production and marketing, among others, which show typical individual entrepreneurial behavior. In this situation, these units would be better prepared to receive the "initial push" advocated by Veiga (2001) cited by Breirenbach and Brandâo (2009). The "initial push" would be characterized by exogenous actions such as technical assistance with rural extension, credit, incubation; preferably.

In the stratum of units with an area in the ">0 to <10.1 ha" range (Table 18), 90.5% (67) of the properties are managed according to the individualist profile and only one unit is in the process of changing. Only 8.1% (6) properties have a non-individualist profile. This situation, historically and culturally, has its origins in Spanish and Portuguese domination, whose chàcaras and/or posteiros were aggregated from the large sesmarias and estâncias during the tropeirismo periods and, later, from the charqueadas. Since the Jesuit-Guarani period, these small properties have supplied labor, subsistence agricultural products, fruit and jams.

Table 18 - Multiple analysis: management profile, estimated income and level of education of head of household for units with an area in the ">0 to < 10.1 ha" range

-	Estimated Income	Level of Education Head of Household
Category n°1 Profile Individualist - PI (67 units)	Category n° 1,"E>R$276.70" + (12) Category n° 2,"D>R$484.97" + (11) Category n° 4,"C1>R$1194.53" + (15) Category n° 3,"C2>R$726.26" + (24) Category n° 5,"B2>R$2012,67" + (5)	Category 1, Illiterate up to 3rd grade Elementary + (29) Category 3, Complete Elementary School + (12) Category 2, 4ª Elementary + (22) Category No. 4, Full Secondary + (3) Category No. 5, Complete Higher Education + (1)
Category n°2 ,PI + IntermediatePII (1 unit)	Category n° 3,"C2>R$726.26" + (1)	Category 2, 4ª Elementary + (1)
Category n° 3, Profile No Individualist - PNI (6 units)	Category n° 5,"B2>R$2012,67" + (1) Category n° 4,"C1>R$1194.53" + (2) Category n° 3,"C2>R$726.26" + (3)	Category No. 5, Higher Education + (1) Category No. 4, Full Secondary + (2) Category 2, 4th Grade + (2) Category 3, Complete Elementary School + (1)
Total	Category n° 3,"C2>R$726.26" (28)	Category No. 1, Illiterate up to Grade 3ª

(74 units)	Category n° 4,"C1 >R$1194.53" (17) Category n° 1,"E>R$276.70" (12) Category n° 2,"D>R$484,97" (11) Category n° 5,"B2>R$2012,67" (6)	Elementary (29) Category No. 2, 4ª Elementary School (25) Category 3, Complete Elementary School (13) Category No. 4, Complete High School (5) Category No. 5, Higher Education Completed (2)

Source: Prepared by the author.

The excessively precarious economic and educational situation of the farmers remains to this day. Their ethnicity is made up of Indians, Afro-descendants, Spaniards and Portuguese, in short, a mixture of Gaudíes who provided some service and came to form the well-known gaùcho missioneiro. Once the republic was established with the political force, power, money and weapons displayed by the big landowners and political bosses, the people lived off the blessings of the big landowners and their work in rural services and farm kitchens. They were loyal to their bosses, the owners of the large estancias, and when they subverted the order of the powerful, they knew they would suffer violence. They also survived on smuggling, which is historical in the region. And they used the culture of fear and silence to preserve their lives.

With the arrival of Prussian-German settlers at the end of the 19th century, the "whitening" of the population began. Agriculture gained momentum. The arrival of Italian and Polish families in the 20th century reinforced the whitening of the population. The legacy of these episodes in the stratum in question can be seen in the large contingent of rural pensioners (Table 11), who took part in the beginnings, heyday and decline of agricultural cooperatives in the municipality. This explains their aversion to cooperatives and their strong disagreement with sales to cooperatives and collective purchases. From then on, this individualistic profile proliferated in the region, where it is still very expressive in units of up to 10.1 hectares.

In the stratum of units with an area in the "=>10 to <50 ha" range, the presence of the individualist profile is significant, but 18 units are classified as PII, considering all categories of estimated income, all levels of education and no gender distinction (Table 19).

It is very likely that the 18 units with a size of between 10 and 50 ha and an individualistic management profile+ (table 19), as a result of this condition, are more susceptible to undertaking neoendogenous changes and actions. Consequently, the demonstration effect of implementing such alternatives would naturally attract neighbors with an individualistic profile, both in the first and second strata (67 and 105, respectively). And so the local development process would begin.

Table 19 - Multiple analysis: management profile, estimated income and level of education of the head of household for units with an area in the "=>10 to <50 ha" range

Management Profile	Estimated Income	Level of Education Head of Household
Category n°1 ,Individualistic Profile - PI	Category n° 2,"D>R$484.97" + (21)	Category No. 1, Illiterate up to Grade 3ª

(105 units)	Category n° 1,"E>R$276,70"" + (5) Category n° 3,"C2>R$726.26" + (28) Category n° 4,"C1>R$1194.53" + (24) Category n° 5,"B2>R$2012,67" + (22) Category n° 6,"B1>R$3479.36" + (5)	Elementary + (34) Category No. 2, 4ª Elementary School + (41) Category 3, Complete Elementary + (24) Category 5, Complete Higher Education + (3) Category No. 4, Full Secondary + (3)
Category n°2 ,PI + Intermediate - PII (18 units)	Category n° 7,"A2>R$6563.73" + (2) Category n° 6,"B1>R$3479.36" + (3) Category n° 4,"C1>R$1194.53" + (7) Category n° 3,"C2>R$726.26" + (3) Category n° 5,"B2>R$2012,67" + (3)	Category No. 4, Full Secondary + (3) Category 3, Complete Elementary School + (9) Category 2, 4ª Elementary + (5) Category No. 1, Illiterate up to Grade 3 Elementary + (1)
Category 3, Non-Individualistic Profile - PNI (18 units)	Category n° 8,"A1>R$9733.47" + (1) Category n° 7,"A2>R$6563.73" + (3) Category n° 6,"B1>R$3479.36" + (4) Category n° 5,"B2>R$2012,67" + (8) Category n° 4,"C1>R$1194.53" + (2)	Category No. 5, Complete Higher Education + (3) Category No. 4, Complete High School + (5) Category No. 3, Complete Elementary School + (3) Category No. 2, Primary 4 + (4) Category No. 1, Illiterate up to Grade 3 Elementary + (3)
TOTAL (141 units)	Category n° 4,"C1>R$1194.53" (33) Category n° 5,"B2>R$2012,67" (33) Category n° 3,"C2>R$726.26" (31) Category n° 2,"D>R$484,97" (21) Category n° 6,"B1>R$3479.36" (12) Category n° 1,"E>R$276.70" (5) Category n° 7,"A2>R$6563.73" (5) Category n° 8,"A1>R$9733.47" (1)	Category No. 2, Primary 4 (50) Category No. 1, Illiterate up to Grade 3 Elementary (38) Category 3, Complete Elementary School (36) Category No. 4, Complete High School (11) Category No. 5, Complete Higher Education (6)

Source: Author's elaboration.

Table 19 shows that the PII and PNI categories are made up of 18 units each. In the PNI group, compared to the PII group, there is a predominance of units in the higher income and higher education strata.

In the stratum of units with an area in the "=>50.1 to <100.1 ha" range (Table 20), the individualist profile continues to predominate and the phenomenon observed in the previous paragraph regarding the income and education strata is repeated.

Table 20 - Multiple analysis: management profile, estimated income and level of education of the head of household for units with an area in the "=>50.1 to <100.1 ha" range

Management Profile	Estimated Income	Level of Education Head of Household
Category n°1 ,Individualistic	Category n° 5,"B2>R$2012,67" + (5)	Category No. 1, Illiterate up to Grade 3ª

Profile - PI (24 units)	Category n° 2,"D>R$484.97" + (1) Category n° 3,"C2>R$726.26" + (1) Category n° 4,"C1>R$1194.53" + (9) Category n° 6,"B1>R$3479.36" + (5) Category n° 7,"A2>R$6563.73" + (1) Category n° 8,"A1>R$9733.47" + (2)	Elementary + (3) Category 3, Complete Elementary School + (10) Category No. 2, 4ª Elementary School + (8) ' Category No. 4, Full Secondary + (2) Category No. 5, Higher Education + (1)
Category n°2 ,PI+ Intermediario - PII (04 units)	Category n° 8,"A1>R$9733.47" + (2) Category n° 6,"B1>R$3479.36" + (1) Category n° 4,"C1>R$1194.53" + (1)	Category 2, Primary 4ª + (2) Category 3, Complete Primary + (2)
Category 3, Non-Individualistic Profile - PNI (05 units)	Category n° 7,"A2>R$6563.73" + (1) Category n° 8,"A1>R$9733.47" + (2) Category n° 6,"B1>R$3479.36" + (2)	Category No. 5, Complete Higher Education + (2) Category No. 4, Full Secondary + (3)
TOTAL (33 units)	Category n° 4,"C1>R$1194.53" (10) Category n° 6,"B1>R$3479.36" (8) Category n° 8,"A1>R$9733.47" (6) Category n° 5,"B2>R$2012,67" (5) Category n° 7,"A2>R$6563.73" (2) Category n° 2,"D>R$484,97" (1) Category n° 3,"C2>R$726.26" (1)	Category 3, Complete Elementary School (12) Category No. 2, Primary 4 (10) Category No. 4, Full Secondary (5) Category 1, Illiterate up to 3rd grade Elementary (3) Category No. 5, Complete Higher Education (3)

Source: Author's elaboration.

In Table 20, the individualist profile prevails in units over 50 ha up to 100 ha.

Again, the individualistic profile prevails in table 21.

Table 21 - Multiple analysis: management profile, estimated income and level of education of the head of household for units with an area in the ">0 to <100.1ha" range

Management Profile	Estimated Income	Level of Education Head of Household
Category n°1 ,Individualistic Profile - PI (194 units)	Category n° 1,"E>R$276.70" + (17) Category No. 2,"D>R$484.97" + (33) Category No. 3,"C2>R$726.26" + (52) Category No. 4,"C1>R$1194.53" + (46) Category No. 5,"B2>R$2012.67" + (33) Category No. 6,"B1>R$3479.36" + (10) Category No. 8,"A1>R$9733.47" + (2)	Category 1, Illiterate up to 3rd grade Elementary + (66) Category 2, 4th Grade + (70) Category 3, Complete Elementary School + (45) Category No. 5, Complete Higher Education + (5) Category No. 4, Full Secondary + (8)

	Category n° 7,"A2>R$6563.73" + (1)	
Category n°2 ,PI+ Intermediate - PII (23 units)	Category n° 7,"A2>R$6563.73" + (2) Category n° 8,"A1>R$9733.47" + (2) Category n° 6,"B1>R$3479.36" + (4) Category n° 4,"C1>R$1194.53" + (8) Category n° 3,"C2>R$726.26" + (4) Category n° 5,"B2>R$2012,67" + (3)	Category 3, Complete Elementary School + (11) Category No. 4, Full Secondary + (3) Category 2, 4ª Elementary + (8) Category No. 1, Illiterate up to 3rd grade + (1)
Category 3, Non-Individualistic Profile - PNI (31 units)	Category n° 7,"A2>R$6563.73" + (4) Category n° 8,"A1>R$9733.47" + (4) Category n° 6,"B1>R$3479.36" + (7) Category n° 5,"B2>R$2012,67" + (9) Category n° 4,"C1>R$1194.53" + (4) Category n° 3,"C2>R$726.26" + (3)	Category No. 5, Complete Higher Education + (6) Category No. 4, Complete Secondary Education + (11) Category No. 2, Primary 4ª + (6) Category No. 3, Complete Primary Education + (4) Category No. 1, Illiterate up to Primary 3 + (4)
TOTAL (248 units)	Category n° 3,"C2>R$726.26" (59) Category n° 4,"C1>R$1194.53" (58) Category n° 5,"B2>R$2012,67" (45) Category n° 2,"D>R$484.97" (33) Category n° 6,"B1>R$3479.36" (21) Category n° 1,"E>R$276.70" (17) Category n° 8,"A1>R$9733.47" (8) Category n° 7,"A2>R$6563.73" (7)	Category No. 2, Primary 4 (84) Category 1, Illiterate up to 3rd grade Elementary (71) Category 3, Complete Elementary School (60) Category No. 4, Medium Full (22) Category No. 5, Higher Education Completed (11)

Source: Author's elaboration.

In summary, in all the area strata and for all the properties, the proportion of units with individualistic management is never less than 72.73% of the total in each stratum (Table 22). In the consolidation of all 248 properties (Table 22), it can be concluded that the individualist profile (PI) is expressively prevalent at 78.23%.

Table 22 - Composition of management profiles by area strata, in all the properties surveyed

Area strata (in ha)	PI	PII	PNI	Properties
>0 a <10,1	**90,54%**	1,35%	8,11%	**74**
=>10 a <50,1	**74,48%**	12,76%	12,76%	**141**
=>50,1 a <100,1	**72,73%**	12,12%	15,15%	**33**
>0 a <100,1	**78,23%**	9,27%	12,50%	**248**

Source: Author's elaboration

3.4 ANALYSIS OF THE INFLUENCE OF HUMAN AND SOCIAL CAPACITY ON LOCAL DEVELOPMENT - SAO BORJA- RS.

An analysis of the survey data is presented, where the indices of trust, leadership, entrepreneurship and schooling are plotted for the municipality of Sao Borja. This shows the data for the municipality of Sao Borja: trust 0.39, leadership 0.33, entrepreneurship 0.36 and schooling 0.52.

In an analysis of the data from Sao Borja, the respondents explain that the confidence index is considered insufficient. However, the creation of UNIPAMPA and the Farroupilha Federal Institute (IFF) in Sao Borja, both from 2008 and classes in 2010, as a positive factor in education, increased the index. The index was no longer affected by the desire of young and middle-aged people to leave the municipality due to lack of employment and opportunities to improve their income. In the view of the respondents, there is too much talk and not enough action. They say they want less promises and more action. Of those surveyed (39), they don't trust the municipality to develop in the short term. Respondent 74 makes a constructive criticism: "Education is lacking. Respect is the basis of education. Human beings have to challenge themselves! ".

The respondents considered the leadership index to be insufficient, and commented that in general politicians act to promote themselves, for their own interests, and do little for the community. There is a lack of unity, party squabbles and clientelism. As for the economic leaders, they comment that they do the minimum for the community, occasionally citing businessmen as a positive example. Respondent 74 says: "Politicians lack responsibility. It's the people who have to do it. Voters don't hold politicians accountable, there's a lack of attitude and they don't hold politicians accountable".

As for the entrepreneurship index, the respondents rated it as insufficient. They explain it partly by their "cultural" nature, their "hard-earned" origins - mestizo and indigenous blood; which would characterize "complacency" or "passivity". At the very least, there is "attitude" and "perseverance", while there is "fear of investing" or "laziness". The entrepreneurs who do exist do so out of a need to survive. They point out the difficulty of starting a business because they need start-up capital, and also refer to the fact that it is not easy to start a business in terms of regulations, taxes and fees. Companies open and close. As for family farmers, they lack the courage to undertake. The "outsiders" are more entrepreneurial, the Germans and Italians. Agricultural cooperatives are in decline.

Sao Borja has 6.21 years of schooling. In relation to the difference between the lowest and highest levels of schooling in the municipalities of Rio Grande do Sul, this results in a schooling index of 0.52.

The final result is that of Sâo Borja, with a ratio of 0.40 to 1.0.

We can see the municipality's need to evolve and develop in the aspects of this research, which are: trust, leadership, entrepreneurship and schooling. The results in the municipality represent an index considered insufficient, 0.40. There is a lot to be done by São Borjense society and the challenge is to increase continuing education, to increase confidence, the participation of leaders in the community, increase individual and collective entrepreneurship; and cyclically increase schooling.

3.5 POSSIBILITIES FOR NEO-ENDOGENOUS DEVELOPMENT IN THE REGION, BASED ON THE THEORETICAL, EMPIRICAL AND SITUATIONAL ANALYSES CARRIED OUT

This subsection presents an analysis of each of the main aspects (categories and subcategories) that should be considered in neo-endogenous development projects identified in Table 8, associated with their framing in the terms: strengths, weaknesses, threats and opportunities. The use of these terms is exclusively aimed at creating a standardized nomenclature in order to make it easier to understand the indications of intervention alternatives:

Human/Cultural Capital Category -

1.1) Individuals and families

- Declining population. Significant number of pensioners and people on sickness benefit (64.23%). Exodus rural excessive. Low attractiveness for young people to remain in rural areas (Weak Point).

1.2) Intellect - Capacities-

- Conceptual-Ethical Skills

- Reasonable ability to achieve goals and objectives. Perceptions of ethics. Considering the prevailing individualistic culture, it is necessary to periodically recycle the principles of living together as a community. Not an impediment to rural development;

- Human Skills - Communication, Motivation, Leadership, Teamwork, Empowerment

- These capacities need to be qualified and developed, especially with regard to young people (Weak Point);

- Technical Skills - (Having the knowledge) Knowledge. Doing concrete and practical things.

- These skills also need training, qualification and development. There is a strong tendency towards resistance and skepticism. Insufficient opportunities for practical training and technical support, as only 11.3% of farmers have technical assistance (Weak Point);

- Management Skills - Planning, Organizing, Directing, Controlling.

- Insufficient number of owners with managerial skills. In general, they have poor or no management skills (62.5% have studied up to 4th grade). This weakness can be reversed with continuing education, interactions and institutional partnerships in training activities; and

- Attitude - (Wanting to do) Entrepreneurial flair, risk-taking, innovation.

- Historically, this aspect has been suppressed by power and politics. Its reversal depends on continuing education, training for entrepreneurship, the availability of incentives and initiatives for innovation. Individual empowerment must be exercised with freedom and decision-making autonomy

(Weak Point).

I.3) Continuing education

- Great weakness.

- The level of education of the heads of household is far too low (62.5% have studied up to 4th grade; and the schooling-education index, obtained in the assessment of human and social capacity, was only 0.52);

An alternative is presented:

a) Rural Family House (2012). A recent initiative to address this issue is the Rural Family House (CFR), which is part of the National Program to Strengthen Family Farming - PRONAF. It is a place within a municipality or region for the technical, human and managerial training of young people from rural and fishing areas. The House allows people to train, adapt and evolve in their profession, together with their family and the community in which they live. It aims to encourage young people to stay in their own region, creating alternatives for work and income, from the perspective of the Solidarity Economy. The professional courses offered, under the Alternance Pedagogy approach, integrate general training (interdisciplinarity), social and human education and the development of a teamwork spirit. A CFR is being set up in the neighboring municipality of Santo Antônio das Missoes, with classes due to start in 2014. It's a good reference point for São Borja to also claim its home.

II) Share Capital

- It is currently a weak point and a major challenge. However, it has great potential, provided that the tradition of collective work that existed in ancestral indigenous communities is revived.

11.1)Actors in communities

- The individualistic culture hinders the possible experiences of the actors in articulating access to institutional, financial, operational, marketing and logistical resources. The individualistic culture is present and ingrained in the community and among rural producers. Local businesses are inert when it comes to any kind of action in this direction. The settlements have difficulties in terms of cooperation and trust, but there are already groups organizing themselves through initiatives based on the solidarity economy and fairs with the participation of the producers and the Catholic Church of Bairro do Passo. Local politicians don't see this as a priority. There are associations that exist on paper, but are not active, and there are also associations such as Nhupoleite, which organizes and carries out actions to improve equipment, etc. There is a lack of integration between rural producers and political, institutional and economic power. It is a huge challenge to overcome these obstacles and make an effective contribution to rural development.

11.2)Cultural diversity

- The region is very rich and lacks value and opportunities;

- Ethnicities

- The indigenous population predominates in the region. Indigenous people are still relegated as dispossessed and treated with discrimination, a situation also faced by Afro-descendants. Citizenship initiatives generate opportunities and actions have been taken by the federal government. However, they are still a long way from including these populations with dignity, through the generation of work and income.

- Gender

- The female gender, historically recognized as "the fighters" in rural areas, despite the influence of Gaucho machismo, has strong entrepreneurial potential. Women have enhanced human skills and are already taking on managerial tasks on farms. They should be encouraged with continuing education, training and capacity building. They represent a great opportunity for rural development;

- Local knowledge

- This aspect is not highly valued in the region. However, at the end of 2012, the Committee for the Celebrations of the 330th anniversary of the establishment of the municipality of Sâo Borja recommended the "Origona or Origone Cheese" project, presented by the author (Figure 6) and the 330th anniversary committee's recommendation is included in Appendix B. This project is an opportunity to revive a cheese from the time of Spanish/Portuguese domination and the tropeirismo in the territory of Sâo Borja. It is made from peaches, roasted in the sun, sweetened and turned into cheese without refrigeration. Its implementation will provide certification of regional origin, according to Herter (2011), for family farmers in São Borja.

Figure 6 - Origona or Origone cheese project proposal

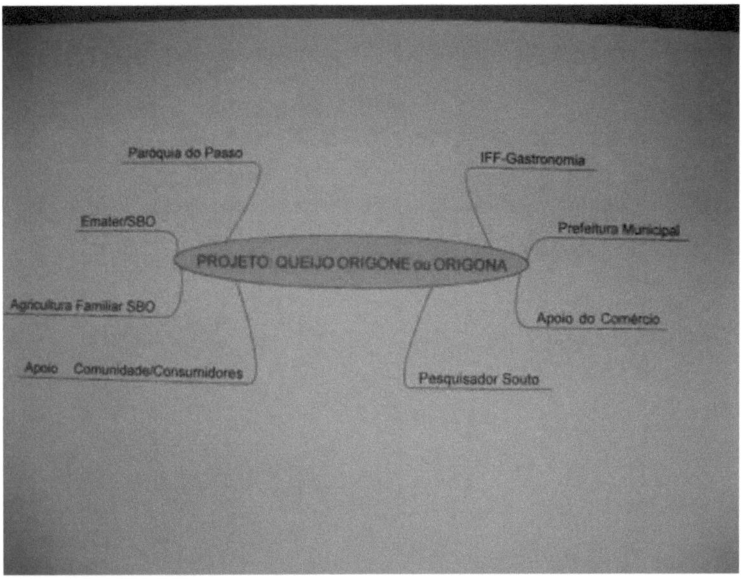

Source: Author's elaboration.

11.3) Confidence

- Only 34.7% of farmers are members of cooperatives or producer unions, and the trust index in the city was 0.39. Under these conditions, the level of trust is very precarious, characterizing a weak point. Reversing this performance will require actions to improve education, increase the level of entrepreneurship and promote local leadership.

11.4) Partnerships and Networks - Incipient and punctual. In this analysis, it is determined as a weak point. However, suggestions have been made for models that are suitable for neo-endogenous development in the region:

a) Designing a local development program. The model proposed by Birkholzer (2005) from the Technical University of Berlin presents a concept for a strategic local economic development program. The model describes local economic development as a process (not a result), which takes time and patience. This model can be useful for an organization or a community to identify where it has already reached, what is missing and what needs to be done next. It involves the development process in three phases:

- The first phase of community building (stages 1 to 4): the first stage corresponds to the analysis of local economic and social structures; the second, the popular planning process involving all the people concerned; the third, building decentralized support and promotion instruments and the fourth, encouraging decentralized networks;
- Second phase of community development (stages 5 to 7): the fifth stage, guidance, education

and information to assume economic solidarity; the sixth, the promotion of public centers for project development and innovation; the seventh, social marketing and/or new relationships between producers and consumers;

- The third stage establishes a functional community economy (stages 8 and 9): the eighth stage is the promotion of new types of social or community-oriented enterprises; and the last stage, the ninth, solidarity finance and/or alternative socially responsible financial instruments.

Figure 7: Design of a local economic development program

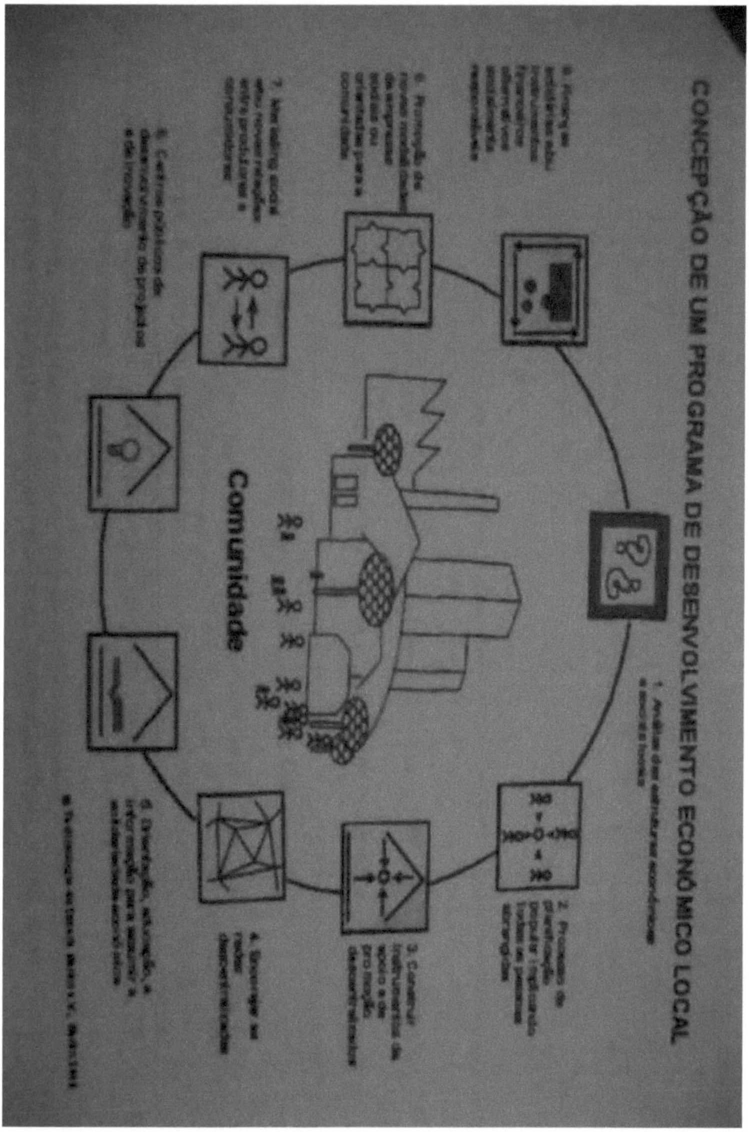

Source: Birkholzer (2005), photo by the author.

b) Cooper Jeito Caseiro - Sao Francisco de Assis (Cooper Jeito Caseiro, 2012). According to the agricultural technician, the Jeito Caseiro Cooperative has 33 members and a capital quota of R$ 50.00 and a monthly fee of R$ 12.00.

With a grant from the Sao Francisco de Assis city council, it aims to: meet local demand for food, generate income at farm level, add value to production, organize the productive sector, promote local trade, maintain and succeed the property, focus on agro-industry and accompany the producer.

The program's objectives are:

- Meeting local food demand;
- Generate income at the property level;
- Adding value to production;
- Organization of the productive sector;
- Promote local commerce;
- Property maintenance and **succession**;
- Focus on agro-industry;
- Accompany the producer.

The factors taken into account in internal and external analysis:

- Formation of partnerships (Emater, Trade Unions, Association of Bank Branches);
- Clear objectives;
- Dedication/involvement;
- **Permanent Construction** Program (long-term project);
- Forming new productive habits;
- Market conquest (involves money);
- Competition (can be a positive factor);
- Political issues (partisan division/political favoritism);
- Training/qualification (Fundamental);
- Entrepreneurial vision (where you want to go);
- Cooperative spirit;
- Participation by all sectors of society (Example: City Hall transfers R$1,500.00 per month to the cooperative);
- Agro-industries: cookies, cheeses, strawberries, colonial eggs, homemade pasta.

The suggested models are references for the municipality of Sao Borja and can be adopted as Benchmarking ;[17]

---

17 Benchmarking - In free translation, as the **best practices** that can serve as a model and adapted to the socio-cultural conditions of the people of Sao Borja, as long as they change their individualistic culture.

11.5) Entrepreneurship (individual and collective) -

- It exists in the region, but with poor performance. In the lexical analysis, it appeared insistently linked to the nature of "culture", "complacency" or "passivity", with a minimum of "attitude", "perseverance" and, in general, "fear of investing". The Socio-Productive Situation Survey identified entrepreneurial landowners and landowners known as rural entrepreneurs. However, their index did not exceed 0.36 when measuring urban human and social capacity.

- It needs to be fostered mainly with training activities, through informal partnerships which are great opportunities for alternatives for local development. For example, a settler in Sao Miguel has a dam, but he has health problems. So he gives it to his neighbors who don't have water for irrigation, in a process of payment for inputs and products. This is resulting in the installation of a shared, irrigated vegetable garden (Figure 8).

Figure 8 - Settler collective entrepreneur. Cambuchim, Sao Miguel locality

Source: Photo by the author (2012).

11.6) Associativism and Cooperativism -

- Considered in the analysis as a threat to rural development in the region. Plotted in the field research, which showed the decline of agricultural cooperatives in the region. An aversion to agricultural cooperatives among rural landowners. In order to reverse the skepticism towards cooperativism, there is a need for cooperative education, training for cooperative managers, active participation by members and alternating leaderships. The process of revitalizing cooperativism culminates with professional management, communication and transparency in the management of the cooperative. A well-managed cooperative induces rural development and increases competitiveness.

- In the stratum of up to 10 hectares, this corroborates Rosa (2011) and Porto (2007). It is suggested that family farming in Sao Borja needs to break away from its individualistic profile and dare to innovate to solve problems, but also to act cooperatively. Breaking out of the vicious circle requires neo-endogenous solutions;

III) Political Capital -

- It is under the control of leaders and parties, as well as people's ability to influence political decisions.

111.1) Power structure

- Relations between the municipal, state and federal governments are dispersed and non-interactive, characterizing this aspect as a weak point for rural development.

111.2) Political leadership

- In the investigation of human and social capacity, respondents gave leadership the lowest score (0.33). In the lexical analysis, the most numerous indications were that local politicians act to "promote themselves" and "do little for the community", with "party fights" and "clientelism". There were also many statements agreeing with the other analyses, and in Table 10 this aspect is in second place in the *ranking* of citations (caudilhismo, coronelismo). Under these conditions, political leadership is a weak point.

111.3) Political participation

- The survey of the socio-productive situation revealed that the income generated is insufficient and that farmers have to leave home early to take the bus, often in the rain or frost, to demand political action for the community. These situations, when too frequent, denigrate and hinder political participation and harm rural activity. This, combined with bureaucracy and a lack of understanding of political processes and procedures, helps to discourage rural landowners from participating in politics. It also emerged that "Voters don't hold politicians to account, they lack the attitude to hold politicians to account and they don't hold politicians to account". This is therefore a weak point.

111.4) Empowerment

- According to Pereira (2006), empowerment begins individually and becomes collective. In this context, empowerment is a weak point. In order to reverse this situation, it is necessary to implement individual and collective actions aimed at decision-making freedom and autonomy.

IV) Financial Resources -

- They limit operational activities and are fundamental to the financial survival of rural properties. In the study, they can be considered as opportunities, threats and weaknesses.

IV.1) External resources

- Non-repayable

- It is essential in the sense of the "initial push" advocated by Veiga (2001), i.e. for the realization of a neo-endogenous project. It represents an important opportunity, provided it is available.

- Credit availability

- A lot of financial resources have been and continue to be made available to the region. However, the crucial problems of rural development also involve resolving debts, payment capacity, agricultural insurance actions and payment settlements, all of which would generate institutional trust. Access to bank procedures and the fees charged sometimes delay credit, postponing investment and disrupting the synchronization needed for rural activities. This highlights the need for personal and farm financial planning using appropriate methods. It should also be noted that this is linked to managerial skills (Item I.2 Intellect-Skills). In this study, this aspect is characterized as a threat. However, it could be an opportunity under the neoendogenous approach.

IV .2) Internal resources/savings

- Considering that 67.3% of families earn incomes of up to R$1,194.53; that 64.13% of farmers supplement their income with pensions and sickness benefits; that 84.2% have practically no conditions of their own to invest; and that the level of indebtedness is a threat to increased competitiveness - this is undoubtedly a weak point.

V) Physical and Material Resources -

VI 1) Earth

- The land is considered a strong point. The region allows for a variety of crops, such as fruit growing, which has been recognized since the missionary period, viticulture, beef and dairy farming, sheep farming, pig farming, olive growing, etc.

V.2) Water

- It is scarce for livestock on 41.5% of the properties and non-existent for crops on 61.3% of them. This can be seen as a threat to the worsening of the current socio-productive situation, and as a weak point for the development of new projects. The region has long lived with the recurring climatic effects of drought/droughts. However, the territory of São Borja is very privileged in terms of rivers, streams and, above all, dams and reservoirs. There are some small properties with dams capable of irrigating one or even two hectares. In this aspect, there is a valuable opportunity.

V.3) Electricity

- It's not a problem. It exists in 99.6% of properties. It's a strong point. However, in the vast majority of properties, the power supply is single phase, with insufficient capacity to drive the motors needed for irrigation. So, in this respect, it becomes a weakness.

V.4) Inputs

- In 37.5% of the properties, the use of inputs is low to regular (Weak Point). However, organic production is already significant in the case of vegetables and subsistence production (cassava, sweet potatoes, etc.). However, due to a lack of organization, initiative and qualification, low-income rural producers are missing out on greater opportunities due to the lack of product certification that would enable them to access this important and rapidly developing market niche.

V.5) Roads

- The region is served by three federal highways, which are in a regular state of repair, and dirt roads, which make it difficult to transport produce, especially during rainy periods. The biggest obstacle to development is the enormous distance from the major consumer centers (Weak Point).

V.6) Livestock

- It's an acceptable name for animals in production. They exist in the region, but with unimproved genetics (Weak Point).

V.7) Infrastructure

- Availability of sheds, buildings, structure for production activities, etc. (Weak Point).

V.8) Machinery and Equipment

- Only 19% of the owners have their own machinery; 11.7% use it in partnership; and 27.8% use the community agricultural patrol. Therefore, it is clear that the farms are not very technologically advanced (Weak Point).

VI) Resources from Operating Activities -

VI.1) Products

- Production is poorly diversified. The main threats to competitiveness are drought, lack of income, aerial herbicides, lack of marketing, cattle rustling and debt (Weaknesses). However, there are unique climatic influences and potential competitive advantages, such as the number of hours of sun exposure and varying temperatures, which broaden the opportunities for production.

VI.2) Processes

- Producers have poor qualifications and management skills and the level of technification is low (Weak Point).

VII) Marketing resources

- Only 3.2% of farmers said they were satisfied with marketing.

VII.1) Local buyers

- They should be encouraged. The scale of production is low, there is a lack of political will and interest in creating alternatives to boost it (Weak Point).

VII.2) External buyers

- The challenge of winning over foreign buyers involves partnerships with the "S" system (Sebrae, Senar), with a view to prospecting for buyer markets, right from the stage of drawing up the neo-endogenous project (Opportunity).

VII.3) Contracts

- The challenge of winning over foreign buyers involves partnerships with the "S" system (Sebrae, Senar), with a view to prospecting for buyer markets, right from the stage of drawing up the neo-endogenous project (Opportunity).

VII.4) Consumption habits

- Where they don't exist, they should be fostered with the help of and/or through partners (Opportunity).

VIII) Institutional Resources

VIII.1) Technical assistance and rural extension

- Under the responsibility of EMATER/ASCAR, which has been providing services in rural areas for decades, and Coptec, which offers social assistance in rural settlements. Due to the scarce human and financial resources available to these entities, their actions are too precarious, despite the proven qualifications of their technicians. Only 11.3% of landowners reported using technical assistance (Weak Point).

VIII.2) Educational institutions

- In Sao Borja, UNIPAMPA (Federal University of the Pampa), the Farroupilha Federal Institute and UERGS (State University of Rio Grande do Sul) predominate in face-to-face education. Given the precariousness of the region's small and medium-sized rural properties, the opportunities for partnerships and contributions from these higher education institutions are promising (Strong Point).

VIII.3) Government bodies

- The municipal, state and federal bodies present in the region are unable to coordinate their actions efficiently (Weak Point).

VIII.4) Non-governmental organizations

- They find it difficult to act due to the barriers imposed by the current individualistic culture (Weak Point).

IX) Logistical resources

IX.1) Storage and stockpiling

- It requires long-term investment, due to the poverty of the region and, consequently, the restriction of the owners' ability to pay (Threat and Weakness).

IX.2) Internal transport

- The dirt roads make transportation difficult and increase dependence on outsourcing or own cars for transportation, whose old vehicles frequently break down. The situation is poor and precarious (Threat and Weakness).

IX.3) External transport

- Incipient. Without a scale of production, external transportation is punctual and inexpressive (Weak Point).

X) Technological/Informational Resources

X.1) Technology

- Perceived as an exception in the few properties that own and use it (Threat and Weakness).

X.2) Informational

- It is perceived as an exception in the few properties that have it and use it. The incipient rural internet is a threat to competitiveness and a weakness.

The quantification of the terms for the aspects investigated, in terms of the possibilities for neo-endogenous development in the region, shows the following distribution:

- Weaknesses - 32 records;
- Opportunities - 11 records;
- Threats - 8 records; and
- Strong point - 3 records.

Faced with such an unfavorable situation, the big challenge is to come up with an intervention proposal in which the endogenous and exogenous factors, characterized as weaknesses and threats, are converted into strengths and opportunities, in an integrated and shared production system supported by public and private partnerships.

3.6 PROPOSAL FOR AN INTEGRATED MULTI-FOCUS PILOT PROJECT WITH PRODUCTIVE ACTIVITIES FOR NEO-ENDOGENOUS DEVELOPMENT IN THE SÂO BORJA - RS REGION

The 248 small and medium-sized farms sampled in the survey are spread over 34 locations[18] , occupying a total area of almost 6,000 hectares. 923 people live and work on them.

The basic elements used in the development of this model were taken from the project entitled "Integrating fish farming with fertirrigated fruit farming on small rural properties in the Jaguari Valley/RS" (TUSI DA SILVEIRA et al...), 2010), which was awarded 2nd place in the first edition of

---

18 The outskirts of Sâo Borja, Rincâo do Ivai, Mato Grande, Santos Reis, Sâo Ramâo, Rincâo de Sâo Bento, Rincâo da Estiva, Sâo Pedro do Butui, Rincâo da Cria, Sâo Marcos, Rincâo dos Batistas, Sâo Joâo Mirin, Triunfo, Conde de Porto Alegre, Mercedes, Sâo Miguel, Timbaùva, Sambura, Passo da Barca, Sâo Matheus, Sâo Miguel, Açoita Cavalo, Santo Ignacio, Caçacâ, Nhu Porâ, Rincâo do Meio, Rincâo de Santana, Olavo Viana 101, Lajeadinho, Rincâo da Conceiçâo, Vila Brites, Rincâo Sâo Joâo, Encruzilhada, Sarandi.

the "National Regional Development Award 2010: Tribute to Celso Furtado", promoted by the Ministry of National Integration (MI), through the Secretariat for Regional Development Policies (SDR); and also from the "Integrated Family Farming" project (TUSI DA SILVEIRA et al., 2012) submitted to the Petrobras Development & Citizenship Program.

Here are the considerations:

Recruitment for the neoendogenous pilot project for integrated production activities took place through radio announcements and a partnership with the Rural Workers' Union (which also represents family farming in Sao Borja), with a view to motivational meetings and publicizing the project.

Scheduled meetings and technical visits to rural family farmers who are candidates for the neo-endogenous enterprise. Ownership is a prerequisite.

During the visits, the properties with the best soil, relief and water resources will be selected.

By jointly identifying the ability of the rural owner to have a non-individualistic profile, and to encourage collective entrepreneurship and cooperativism in the future.

Carrying out the courses, targeting the best properties/owners who wish to join the neoendogenous project, which will be formalized by prioritizing producers with sufficient attendance and demonstrated attitude.

Subsequently, the technical surveys will be carried out (topography, soil and water analysis, among others), which are necessary for locating and sizing the fields, the integrated production systems and the vocations identified in the previous stages. The focus of irrigation is to promote productive activities, with efficiency and income generation to serve the client market.

Table 23 shows the partnerships that are essential to the project.

Table 23 - Essential partnerships for the pilot project

Partner's name	**Type of contribution (financial, technical, human resources or other)**
Federal and state governments	Non-repayable grants to install and equip the project headquarters, set up dams and orchards, etc. (this would be the "initial push")
Regional Council of Development of the Western Frontier	Institutional support
Sâo Borja City Hall, RS	Institutional support, human and material resources (cessation of machine operators and the equipment needed to restore dams and build dugout tanks, such as backhoes and trucks, among others)

Rural Workers' Union	Institutional support in disseminating the results of the project and in selecting and monitoring rural producers with the necessary profile
Technical Assistance and Rural Extension Company, RS - Associação Sulina de Crédito e Assistência Rural (EMATER/ASCAR)	Technical assistance to new entrants to request resources and implement the system on their properties
Brazilian Micro and Small Business Support Service (SEBRAE), National Rural Learning Service (SENAR)	Technical assistance to new entrants to request resources and implement the system on their properties.

Source: Author's elaboration.

Farmers will receive sufficient technical assistance with rural extension in all the activities of the integrated projects. Technical visits, when necessary, will take place on field days, giving preference to properties whose location makes it easier to meet a greater number of producers, including those who are not committed to integrated neo-endogenous projects.

Financial viability will come from partnerships with the project's submission to federal government entities - the Ministry of National and State Integration - state development agencies, as well as proposals to international regional and local development organizations. External financial support will be sought until the fourth year of the neo-endogenous project, or when the properties obtain a minimum annual gross income of R$ 27,000.00, the beginning of sustainability, as confirmed by the survey of socio-productive profiles in Sao Borja.

The products obtained during the first 4 years of implementation of the integrated projects must be marketed in-natura and, preferably, for the municipality's school meals, in accordance with Law No. 11.947/2009.[19]

Given the need to comply with the 30% requirement for family farming products under federal legislation, according to information from Emater-Sao Borja, in 2012 the percentage of supply did not reach 5%. This creates a market opportunity for family farming in Sao Borja to supply school meals.

With the exception of activities such as fish farming and rabbit farming, provided they are built in accordance with environmental and technical standards, processing facilities or agro-industries necessary for rural development.

With integration in mind, we can mention worm farming, which can be used to sell bait to fishmongers

19 Law N0 11.947/2009 - Determines the use of at least 30% of the resources passed on by the National Education Development Fund (FNDE) for school meals, in the purchase of products from family farming and rural family entrepreneurs, or their organizations, giving priority to agrarian reform settlements, traditional indigenous communities and quilombolas according to article 14. And amended by Resolution No. 25 that the sales limit per DAP per year is increased to R$ 20 thousand. Available: <http://www.mda.gov.br/portal/saf/programas//alimentacaoescolar>, access: May 2013.

and to form the humus and substrate needed for organic fertilization in olive and fruit farming. It also makes use of poultry, pig and dairy cattle manure for the integrated system.

Another example is fruit-growing with the mulberry tree, which would provide breakwaters, and the fruit would be used to make homemade jams and cakes; at the same time, the stalks and leaves would be used to sustain animal nutrition.

Dairy farming, sheep farming, pig farming and goat farming make intensive use of rotational grazing with electric fencing. This highlights the integrated management by groups with family vocations and collective entrepreneurship.

In terms of job creation, the integrated projects will tend to keep young people and families with sufficient income so that, at the end of five years, the farms will be self-sustaining.

The project thus contributes to reducing the rural exodus, training for rural work and focusing on the consumer market for rural products, with quality and a minimum initial scale, contributing to local and regional development.

The necessary pre-incubation, with self-managed companies and at least one per category of related activities and the promotion of universities, federal institutes, in partnership with EMATER/ASCAR - RS, Sao Borja City Hall, the Federal and State Governments.

It should be emphasized that learning-by-doing over four years is an additional guarantee for the effective instrumentation of the project's technical and institutional resources.

The aim is to reduce the impact of the rural exodus by providing professional training through own work on the properties, increasing self-esteem and stimulating concrete professional, social, participatory and community opportunities.

The rational use of water, environmental preservation and the reduction of chemical inputs or the use of organic farming will be used to minimize the impacts and emissions of CO_2 on the planet.

The integrated multifocal neoendogenous project also aims to minimize the effects of droughts and dry spells, which in these integrated systems minimize the harmful effects of irrigation on the territory of Sao Borja.

This proposal for a multi-focus pilot project for neo-endogenous productive activities is not intended to enter into theoretical debates about what is endogenous or exogenous in development. In the proposal, both complement each other. So, just as an example, I would cite the endogenous factors of social capital, the trust and collective work necessary for entrepreneurship. And, exogenous, external financial support and technical assistance to the enterprise.

Table 24 below shows the resources, specifications, situation and actions required/possibilities for this neo-endogenous pilot project.

Table 24 - Resources, specifications, situation and actions

needed/possibilities of the neoendogenous pilot project in Sâo Borja

Resources	Specification	Situation	Actions Needed / Possibilities
Human	- Family Farming: - Level of education of the head of the household: - Job opportunities for young people and women: - Ethnic diversity:	- 211 properties; - Low. 62.5% have completed the 4th grade of primary school; - Migration of 57.3%, mostly young people; - Women working exclusively on household chores; - There are the "hardcore", the Afro-descendant and the gringo (Italian, German or Polish);	- Awareness-raising, motivation and recruitment; - Take this into account when planning qualification training; - Training, capacity building and the purpose of development; - Reviving collective work, which is historic, and preferably collective entrepreneurship with a view to cooperativism; - Retrieval of skills in relation to fishing, fruit growing, sheep farming and innovation with other activities; - Assign everyone the most appropriate activity for their ethnicity, local customs, habits and knowledge;
Resources	**Specification**	**Situation**	**Actions Needed / Possibilities**
Technical and managerial	- Technical knowledge: - Management knowledge:	- There is knowledge of tradition. But there is no knowledge of technology; - Poor management skills;	- Training, capacity building and development with the necessary monitoring of the techniques to be implemented; - Training, capacity building and development with monitoring necessary for the management techniques to be implemented. Emphasis on young people and women;
Resources	**Specification**	**Situation**	**Actions Needed / Possibilities**
Financial	- Non-repayable; - Availability of credit; - Internal savings;	- External resources as the initial push according to Veiga (2001); - Lines of credit already exist for family farming, but because of the indebtedness, owners are unable to access	- It is imperative to start the neoendogenous project until the income of the properties stabilizes; - Financial leverage required after the economic break-even

		them despite the low interest rates currently charged; - Almost non-existent;	point of the properties, - Fostering a culture of saving;
Resources	**Specification**	**Situation**	**Actions Needed/Possibilities**
Operational	- Livestock (dairy cattle, sheep, pigs, poultry, etc.); - Infrastructure;	- They do exist in the region, but with low genetic and productive potential; - Poor infrastructure in the region;	- Promote genetic improvement in partnership with government research institutes and private companies; - Long-term investment needed
	- Machinery and Equipment; - Inputs;	- Low level of technology on farms in the region investigated; - Low use of inputs;	timeframe. Encouraging the use of partnerships and collective entrepreneurship. Credit lines can be made available to producers of the neoendogenous project at banks; - Partnerships are necessary. Including the use of the municipal agricultural patrol. And credit lines can be made available; - Encouraging organic farming or minimizing chemical inputs;
Resources	**Specification**	**Situation**	**Actions Needed/Possibilities**
Productive	- Earth: - Water: - Light and Power - Products: - Processes:	- Owners only; - Scarce in general; - Light is available on 99% of properties. The power to irrigate can be limiting; - Low scale, poor marketing; - Processes are poorly organized and synchronized. Processes are poorly planned	- Certificate of ownership; - Technical study and if there is no action, carry out the construction or renovation. It limits the project; - Technical study required for irrigation alternatives; - Proposal for productive categories in integrated neoendogenous systems; - There is a huge need for technical and professional advice. Describing, organizing,

	- Environmental preservation:	and monitored. Few properties have positive examples; - Poor awareness and action for the environmental preservation;	resizing and new procedures are fundamental; - Need to raise awareness, raising awareness, training and empowering people. Community involvement must be encouraged;
Resources	**Specification**	**Situation**	**Actions Needed/Possibilities**
Institutional	- Technical Assistance with Rural Extension: - Teaching and Research Institutions: -Governmental bodies: - Non-Governmental Organizations	- Plotted in the survey as a weak point; - Weak systemic interactions. Isolated actions; - Low interaction. Weak political coordination; - Weak systemic interactions. Isolated actions;	- Encouraging partnerships and demonstrating the need to increase of technical assistance with effective rural extension for the success of the neoendogenous project; - Fostering partnerships and institutional commitment to the neo-endogenous project; - The challenge is to promote coordinated actions for the neo-endogenous project; - Fostering partnerships and institutional commitment to the neo-endogenous project;
Resources	**Specification**	**Situation**	**Actions Needed / Possibilities**
Marketing	- Internal buyers; - External buyers: - Consumer habits and prospecting for opportunities :	- Plotted in the survey as a weak point; - Need for improvement; - Nonexistent;	- Long-term actions. Partnerships would foster the neo-endogenous project; - Encourage partnerships with the "S" system (Senar and Sebrae) to prospect for markets. Seek partnerships and institutional commitment to the neo-endogenous project; - The challenge is to promote coordinated actions for the neo-endogenous project;

Resources	**Specification**	**Situation**	**Actions Needed / Possibilities**
Logistics	- :	- Plotted in the survey as a weak point;	- Long-term actions. Partnerships would foster the neo-endogenous project;
	- Internal roads, Internal transportation:	- Need for improvement;	- Fostering partnerships and institutional commitment to the neo-endogenous project;
	- External transportation:	- Low production and, consequently, low foreign trade;	- The challenge is to promote coordinated actions for the neo-endogenous project;
Resources	**Specification**	**Situation**	**Actions Needed / Possibilities**
Technology and Information	- Technology:	- Plotted in the survey as a weak point;	- Long-term actions. Partnerships would foster the neo-endogenous project;
	- Information:	- Need for improvement;	- The challenge is to promote coordinated institutional actions for the neo-endogenous project;

Source: Author's elaboration.

General comments:

The products on offer are: pumpkin, lettuce, peanut, blackberry, sweet potato, bergamot, beet, cookie, cake, broccoli, goat, tea, chinchilla, chayote, rabbit, jam, lamb, fish leather, cabbage, cauliflower, cuca, adult cattle waste, adult pig waste, calf waste, sweets, herbs, peas, beans, figs, fish fillets, free-range chicken, jam, wool, oranges, suckling pigs, milk, lemons, papaya, manioc, passion fruit, honey, molasses, watermelon, cantaloupe, corn, green corn, earthworms, strawberries, sheep, eggs, bread, fish, skins, cucumbers, pears, peaches, cheese, radishes, rapadura, cabbage, arugula, grain sorghum, grape juice and tomatoes. In this sample, 65 products are proposed, but with 69 income classifications, divided into short-term income, short-term income, medium-term income and long-term income.

Below is table 25, which lists product categories and periodic income.

Table 25- Classification of proposed activities in relation to periodic income for family farming in the Sao Borja neo-endogenous pilot project

Classification of Activities/Income in periodicity[20]	Short term	Short Term	Medium Term	Long term	TOTAL

20 With regard to periodicity of income, i.e. the cycle of the agricultural product becoming income for the producers, the criteria were established as follows: short-term income of up to three months, short-term income of more than three months to six months, medium-term income of six months to one year and long-term income

Agriculture	0	1	2	0	3
Beekeeping	0	0	1	0	1
Poultry farming	1	1	0	0	2
Dairy farming	2	0	1	1	4
Goat farming	1	1	0	0	2
Rabbit farming	0	1	2	0	3
Fruit growing	0	2	2	9	13
Worm farming	1	0	0	0	1
Olericulture	2	13	6	1	22
Sheep farming	0	1	3	0	4
Fish farming	0	0	3	0	3
Pig farming	0	1	0	1	2
Homemade products	8	0	1	0	9
TOTAL	15	21	21	12	69

Source: Prepared by the author using expert criteria, adapted from Rosa (2011).

In order to give visibility to the proposal, the products for family farming were drawn up and listed with a periodic income. The proposal is not exclusionary, products can be added or removed, but the proposal includes the preservation of local and ethnic habits and customs and the inclusion of new products with the possibility of adding value.

The key factors in the neo-endogenous project are a non-individualistic culture, people's attitude, production techniques with integrated systems, a focus on the market, promotion and incubation, implementation, production and marketing on a large scale, attracting new partners and rural producers with a view to reviving agricultural cooperativism; increasing the virtuous spiral.

Products are proposed for family farming in Sao Borja:

- **Short-term income**: lettuce, cookies, cakes, jams, jellies, milk for cattle, milk for goats, worms, eggs, bread, cheese, rapadura, arugula;
- **Short-term income**: pumpkin, beet, broccoli, kid, chayote, rabbit, lamb, cabbage, cauliflower, beans, free-range chicken, suckling pig, watermelon, melon, green corn, strawberry, radish, cucumber, cabbage, tomato;
- **Medium-term income**: peanuts, sweet potatoes, lamb, tea, chinchilla, fish skins, calf skins, herbs, peas, fish fillets, la, manioc, passion fruit, honey, molasses, corn, sheep, chinchilla skins, rabbit skins, fish, grain sorghum;
- **Long-term income**: blackberries, bergamots, teas, waste from adult cattle, waste from adult pigs, figs, oranges, lemons, apples, pears, peaches and grape juice (SOUTO 2013 adapted ROSA 2011).

---

of more than one year.

The need for sustainability in the neo-endogenous project is emphasized, and for this challenge to be met, it is vital that rural income is planned with short-term, medium-term and long-term incomes. These incomes interact and complement each other with the individual characteristics of the properties and the collective entrepreneurship of the project participants.

Justification for the neoendogenous pilot project:

The proposal for a neo-endogenous pilot project is based on the need for the target audience, the 248 farms, to be fostered socially, institutionally and economically through public-private partnerships . Cultural-historical individualism, low production and marketing, poor individual and collective attitudes, place the region in what could be considered a remote development area. They make it difficult for these producers to gain access to continuing education, technical assistance and rural extension, as well as municipal, state and federal public policies, which lack effective interaction and coordination.

Therefore, we need to think and act differently. The exogenous development solutions of the past no longer work, and endogenous development is in its infancy in the region in question.

The proposal for the neo-endogenous pilot project envisages preserving productive activities, innovating in other activities, the necessary incubation of these with the learning of technical knowledge and fostering local social capital. Generating opportunities for women, attracting young people to stay in rural areas, preserving culture and ethnicity, investing in technology and irrigation.

These aspects are fundamental to the neo-endogenous rural enterprise, both socially and economically, meeting the demands of local and global customers. The alternatives presented provide for the analysis of categories and subcategories in a multisectoral and transversal manner.

FINAL CONSIDERATIONS

This dissertation, a professional master's degree in Strategic Management of Organizations, aimed to study the foundations and possibilities of neoendogenous rural development on farms in the municipality of Sao Borja - RS.

The historical analysis, carried out according to content and citation analysis, showed the most frequently referenced themes: lack of income/lack of commercialization, caudillo/colonel culture and the individualist model. This analysis makes it possible to link the power structure in force in the territory and the lack of income to the individualist model that transforms these factors into an individualist culture.

Historically, the small property segment, from the Spanish, Portuguese and Brazilians onwards, has not been a priority for governments and those in power in the past, and has always been relegated to occasional treatment to satisfy the needs of the segment in question. The rural segment studied is strongly influenced by an individualistic culture, and this culture is a negative condition for rural development. This characteristic was also reported in the analysis of Sao Borja's human and social capacity index.

The survey of the socio-productive profiles of small and medium-sized rural properties in Sao Borja, where 248 properties were investigated, revealed the following results: a small proportion of the rural population (10.6%), a high proportion of retirements and sickness benefits (64.13 %), 67.3 % of the properties in categories called poverty by the estimated income, 62.5 % of the heads of household with insufficient education (schooling), poor technification and use of inputs, a low supply of products for marketing, technical assistance

Incipient, weak associative and cooperative behavior, vulnerability to droughts and droughts, 78% of properties/owners with an individualistic profile, being more aggravating in the stratum up to 10 hectares with (90.54%) of owners in this profile, negatively reinforcing the picture, the rural exodus verified by the number of "taperas", i.e. abandoned rural dwellings, and the involution of the number of young people enrolled in rural schools in the municipality.

The research analyzing the influence of human and social capacity in Sao Borja plotted data from the indices: trust 0.39, leadership 0.33, entrepreneurship 0.36, schooling 0.52. Analyzing the total calculated in the index resulted in 0.40 in relation to 1.0. In other words, 40% is the human and social capacity index in Sao Borja, demonstrating insufficient performance when analyzing this criterion based on field research. The community of the territory lacks the necessary improvements for development in the indices surveyed.

The results show that the fundamentals of neo-endogenous development, as set out in the theoretical and empirical frameworks and field research, are not met in the territory of São Borja.

Highlighting this dissatisfaction with the fundamentals of neo-endogenous development, section 3.6

proposes an integrated multi-focus pilot project for productive activities in neo-endogenous development in the Sâo Borja region. The proposal includes initial considerations of stages, criteria, partnerships and clients, sources of investors for financial realization, theoretical framework, and possible gains. A table is drawn up with the key resources, their specifications, the current situation and necessary/possible actions. The resources analyzed in the region of the proposal are human resources, technical/managerial resources, financial resources, operational resources, productive resources, institutional resources, marketing resources, logistical resources and technological/informational resources. 65 products are proposed in 69 income classifications, divided into short-term income, short-term income, medium-term income and long-term income.

The conclusion is that the key factors in the neo-endogenous project are a non-individualistic culture, people's attitude, production techniques with integrated systems, a focus on the market, promotion and incubation, implementation, production and marketing on a large scale, attracting new partners and rural producers with a view to reviving agricultural cooperativism; increasing the virtuous spiral.

The need to seek the sustainability of the neo-endogenous project is emphasized, and for this challenge to be met, it is essential that rural income is planned with short-term, medium-term and long-term incomes. These incomes interact and complement each other with the individual characteristics of the properties and the collective entrepreneurship of the project participants.

With regard to prospects for the region, two scenarios are presented: a negative future for the rural segment studied, if the current forms and models of production are maintained, and a future with a long-term learning process.

The negative future scenario is based on the context of the individualistic culture in the territory and in the segment studied. This is reinforced by the relations of power/domination that are reflected in the segment's low political and population influence, which is not a priority in the public policies of the municipal government, historically, partially in the state and, currently, more so in the federal sphere.

The lack of production and marketing means that the segment has little economic influence. In addition, there is the disunity of producers due to the lack of associations and cooperatives, as well as the incipient technical assistance in rural extension. In this scenario, there is a decrease or weak perpetuation of small rural properties due to the aging of rural producers, the sharing of properties and the low replacement of new producers for small properties.

Another negative and decisive factor is the lack of income, which makes it unattractive for young rural people to stay on farms, so they move to the city or other regions with more opportunities. As a result, the tendency is to concentrate on large properties, by selling the land and, above all, on large corporate properties.

The second future scenario is a learning process with a paradigm shift towards a non-individualistic profile: individual entrepreneurs, collective entrepreneurs, associations and cooperatives; with the start of neo-endogenous promotion. The actors acting as a community, the municipal government fostering political will and greater budget allocations, as well as human resources for the small farm segment, effective interaction with institutional resources (universities, federal institutes, governments, non-governmental organizations) and fostering partnerships with rural producers, to produce the necessary scale with quality, meeting health requirements and preserving the environment. Working in partnership on a neo-endogenous project.

Continuing education, individual and collective entrepreneurship, the revitalization of cooperativism, real technical assistance with rural extension, projects and actions, financial resources initially in the form of non-repayable grants and after the necessary incubation, this system would become self-financing with the purchase and programmed contracts signed with rural producers, as well as the planning of distribution logistics to customers. The necessary support from commerce and for school meals, acquiring the products, banks refinancing, services and agro-industries, in short, the neo-endogenous process would become a long-term achievement. What is needed in this process is people's motivation, committed leadership, the attitude to implement actions, persistence in the pursuit of objectives, as well as quality of life and the sustainability of the system as a whole.

The suggestion is also made to expand research into small and medium-sized properties in the territory of São Borja and in the southern half of Rio Grande do Sul.

REFERENCES

ABIPEME. ***Socioeconomic Classification.*** Abipeme 2008. Available: < http://www.viverbem.fmb.unesp.br/docs/classificacaobrasil.pdf >, accessed: July 26, 2011.

ABRAMOVAY,Ricardo. ***Paradigms of Agrarian Capitalism in Question.*** 2ª Ed. Sao Paulo: HUCITEC,1998.

ALVES, Adilson Francelino. ***From the Design to the Implementation of Sustainable Rural Development Projects:*** *interfaces and negotiations in the Vida na Roça Project (Paranà).* PhD thesis from the Interdisciplinary Postgraduate Program in Human Sciences. Federal University of Santa Catarina. Santa Catarina, 2008.

ANTUNES, Luciano Medici; RIES, Leandro Reneu. ***Agricultural Management.*** 2nd Ed. Guaiba: Agropecuària, 2001.

BAGUET, A. ***Viagem ao Rio Grande do Sul***. Santa Cruz do Sul: EDUNISC. Florianópolis: PARAULA, 1997.

BAKOS, Margaret Marchiori. ***RS:*** *slavery & abolition*. Porto Alegre: Mercado Aberto, 1982.

BARDIN, Laurence. ***Content Analysis***. 2nd Reprint, la Ed.Sao Paulo: Ediçôes 70, 2011.

BAQUERO, Marcello; CREMONESE, Dejalma. ***Regional development, local democracy and social capital.*** Ijui: Ed.Unijui,2008.

BARQUERO, Antonio Vàsquez. ***Endogenous development in times of globalization***. Porto Alegre: UFRGS, 2001.

BECERRA, Alfredo T.; BRAVO, Xavier L. ***Planificación en los espacios rurales espanoles Aplicación del modelo neo-endògeno para un desarrollo sostenible en las comarcas de metodologia LEADER.*** Area of Engineering Projects. University of Almeria - Spain. Article in UCM Magazine, Observatorio Medioambiental, vol.12, p.49-75, ISSN 1139-1987.

Available::<http://revistas.ucm.es/index.php/0BMD/article/view/0BMD0909110049 A>. Accessed: July 9, 2013.

BIRKHOLZER, Karl. ***Local Economic Development and its Potential.*** Interdisciplinary Research Group on Local Economy at the Technical University of Berlin. Seminar "Local Economic Development", organized by the Network of Associations of Local Elected Officials in Southern Europe. Bosnia Herzegovina, Brckoc, on April 14, 2005. Available: <strateg.dev.loc_karll.birk_vp.pdf>. Accessed: Oct, 2011.

BOEIRA, Nelson; GOLIN, Tau. ***EMPIRE.*** Vol. 2. General History of Rio Grande do Sul Collection. Passo Fundo: Méritos, 2006.

. ***IMPÈRIO.*** Vol. 2. General History of Rio Grande do Sul Collection. The Agrarian Structure. Paulo

Afonso Zarth, p.187 to 212. Passo Fundo: Méritos, 2006.

. ***IMPÈRIO.*** Vol. 2. General History of Rio Grande do Sul Collection. Multiculturalism and Ethnic Frontiers Silvio Marcus de Souza Correa, p.257 to 276. Merits. Passo Fundo, 2006.

BREITENBACH, Raquel; BRANDÃO, Janaina Balk. ***Some constraints for rural development:*** *beyond public policies.* V Meeting of Research Groups "Agriculture Regional Development and Socio-Spatial Transformation, UFSM, 2009.

BRUXEL, Arnaldo. ***The Thirty Guarani Peoples.*** University of Caxias do Sul. Sâo Lourenço de Brindes School of Theology. Porto Alegre: Livraria Sulina Editora, 1978.

CAPRA, Fritjof. ***O Ponto de*** Mutaçào*: A ciência, a sociedade e ªcultura emergente.* 25ª Ed. Sâo Paulo: Cultrix, 1982.

RURAL FAMILY HOME. Available:<

http://www.gestaoescolar.diaadia.pr.gov.br/modules/conteudo/conteudo.php?conteu do=185>; access: 13 Dec. 2012.

CASTELLS, Manuel. ***The Network Society.*** 4th Edition. Paz e Terra. 1999.

CESAR, Guilhermino. ***History of Rio Grande do Sul:*** *Colonial Period.* 3rd Ed. Porto Alegre: Martins Livreiro, 2002.

COLUSSI, Eliane Lucia. IX State History Meeting. National History Association. ANPUH-RS. ***Judicial Sources and their Possibilities in Local Power Studies:*** *the crimes of Sâo Borja.* University of Passo Fundo. 2008. Available: http://www.eeh2008.anpuh-rs.org.br/resources/content/anais/1212424780_ARQUIVO_resumoanpuhfontesjudici ais1.pdf , Accessed 28/01/2012.

COSTA, Alessandra. S. M. da; BARROS, Denise. F; MARTINS, Paulo E. M; ***Historical perspective in administration:*** *new objects, new problems, new approaches.* RAE, ISSN 0034-7580. Sâo Paulo, v.50, n°3. Jul/Set 2010. P.288-299.

COOPER, D. R.; SCHINDLER, P. S. ***Research Methods in Administration.*** Sâo Paulo: Bookman, 2001.

CSCJ. ***Sacred Heart of Jesus School sees Sao Borja through its window 1932-1992.*** Sâo Borja: CSCJ, 1992.

DALLABRIDA, Valdir Roque. ***Regional Development****: why do some regions develop and others not?* 1ª Reprint. Santa Cruz do Sul: EDUNISC, 2011.

DREYS, Nicolau. ***Noticia Descritiva da Provincia do Rio Grande do Sul de Sao Pedro do Sul.*** 4th Edition. Porto Alegre: EDIPUCRS, 1990.

FIGUEIREDO, Adma de. Out of Rhythm: education is still an obstacle in rural areas. **Zero Hora.**

Field and Crop Supplement, journalist Nestor Tipa Jùnior. Porto Alegre, October 20, 2012.

SHEET OF SĂO BORJA. ***Cotrisal Special Edition.*** The technical department was created with Cotrisal in 1956 and expanded in 1965. Sao Borja: Folha de Sao Borja, April 11, 1981.

FLÔRES, Joao Rodolpho Amaral. ***The town of "Sâo Francisco de Borja das Missoes" (1834-1887):*** *elements of history and geopolitics in the formation of Brazil's southern limits.* Santa Maria: Ed. da UFSM, 2012.

FLORES, Moacyr. ***A Revoluçao Farroupilha:*** *sintese riograndense*. 4th Ed. Porto Alegre: Ed. UFRGS, 2004.

FREITAS, Décio. ***Missôes:*** *chronicles of a genocide*. 3rd Ed. Porto Alegre: Movimento, 1998.

FREITAS, Osório Tuyuty de Oliveira. ***The Invasion of Sao Borja.*** Porto Alegre: Livraria do Globo, 1935.

FROEHLICH, José Marcos; DIESEL, Vivien (Organizers). ***Rural development:*** *contemporary trends and debates*. Article: Family Farming and Endogenous Rural Development. Sergio Schneider. Pages 15 to 63. Ijui: Ed. Unijui, 2006.

FUKUYAMA, Francis. ***Trust****: the social virtues and the creation of prosperity*. Rio de Janeiro: Rocco, 1996.

GAY, Joao Pedro. ***Paraguayan Invasion on the Brazilian Border of Uruguay***, by Canon Gay; Commented and edited by Major Sousa Docca. State Book Institute, Sao Lourenço dos Brindes School of Theology. Caxias do Sul: University of Caxias do Sul, 1980.

GIMENEZ, Fernando A. P.; TÓFFOLO, Katia da Silva. In: SOUZA, Eda Castro Lucas de; GUIMARAES, Tomàs de Aquino (Org.). **Entrepreneurship beyond the Business Plan.** Article entitled Rural Female Entrepreneurship: Jane Calamity or Scarlett O'hara?, p. 89 -111. Sao Paulo: Atlas, 2005.

GONZAGA, Sergio S.; SANTOS, Jorge Luiz S. dos. **Good Agricultural Practices:** a new look at the competitiveness of cattle farming in Rio Grande do Sul. In: Ezequiel Rodrigues do Valle. Quality control on farms and access to differentiated markets. Pages 11-13. FEDERACITE XVIII. Bagé: Embrapa Pecuària Sul, 2010.

GIL, A.C. ***Como Elaborar Projetos de Pesquisa.*** 3ª Ed. Sao Paulo: Atlas, 1991.

. ***Methods and Techniques of Social Research.*** 4th Ed. Sao Paulo: Atlas, 1994.

IBGE. ***Agricultural Census.*** Sao Borja code 431800. Brasilia: IBGE, 2006. Accessed: Sep, 2011.

. ***Population Census***. Brasilia: IBGE, 2010. Available at: <

http://www.ibge.gov.br/cidadesat/link.php?codmun=431800 >Accessed: May 2012.

IBGE SIDRA. IBGE (2012).

Available: <http://www.sidra.ibge.gov.br/bda/tabela/protabl.asp?c=2986&z=cd&o=16&i=p>, Accessed: October 22, 2012.

IPEA. ***Radar Book.*** Income 2003. Available:< http://www.ipea.gov.br/destaques/livroradar/03.renda.pdf>. Accessed: 02 Oct. 2012.

HERTER, Pedro Luiz. ***Certifications:*** *a solution for a successful future*. Article published in the book: Sustainability as a Competitiveness Factor in Agricultural Systems. Pg. 76 a 79. Federacite XIX, 2011.

JULIEN, Pierre-André. ***Regional Entrepreneurship and the Knowledge Economy.*** Sao Paulo: Saraiva, 2010.

KHANNA, Tarum. ***Billions of Entrepreneurs:*** *how China and India are changing their future (and everyone else's)*. Rio de Janeiro: Elsevier, 2009.

LAGEOLAM. ***Geoenvironmental Atlas of Sao Borja.*** Federal University of Santa Maria, Environmental Geology Laboratory, Sao Borja City Hall. Santa Maria: UFSM, 2007.

LANGER, Protâsio Paulo. ***The Guaranimissioneiros and Luso Colonialism in Southern Brazil.*** Porto Alegre: Martins Livreiro-Editor, 2005.

LAYTANO, Dante de. ***Origin of Rural Property***. São Lourenço de Brindes School of Theology. Martins Livreiro - Editor. Porto Alegre, 1983.

MARCH, James G. ***How Decisions Really Happen:*** *principles of decision making in organizations*. 1st Ed. Sao Paulo: Leopardo, 2010.

MAURER, Rodrigo Ferreira. ***From Necessity to Intrigue: The Case of San Borja in the Face of the Political and Social Processes of the Years 1750 to 1759***. ANPUHRS, X State History Meeting. July/2010. Santa Maria - RS.

MEIRELES, Magali Rezende G.; CENDÓN, Beatriz Valadares. ***Practical Application of the Processes of Content Analysis and Citation Analysis in Articles Related to Artificial Neural Networks.*** DOI:10.5433/1981- 8920.2010v15n2p77. Inf.Inf., Londrina, v.15, n°2, p.77-93, jul./dez.2010.

MDA/DESER. **Final Report of the National ATER Seminar**. Available:< http://www.deser.org.br/publicacoes/Sistematiza%C3%A7%C3%A3o%20final.pdf>, access:10 abr. 2013. Brasilia, 2008.

MORGAN, Gareth. ***Images of the Organization***. Sao Paulo: Atlas, 1996.

O'DONNELL, Fernando O. M. ***Noticias dos Combates:*** *de capão do Mandiju e Estância dos Figueiredos*. Porto Alegre: Martins Livreiro Editores, 1985.

OLIVEIRA, José Roberto de. **Pedido *de Perdao ao Triunfo da Humanidade:*** *a importância dos 160 anos das missoes jesuitico-guarani.* 2ª Ed. Porto Alegre. Martins Livreiro-Editora, 2011a.

OLIVEIRA, José Roberto de. ***Analysis of the Influence of Human and Social Capacity on Local Development: the cases of Caxias do Sul, Feliz and Sao Nicolau in Rio Grande do Sul.*** Master's dissertation in Development.

UNIJUi, 2011b.

PADILHA, Paulo Roberto Paim; TRENTIN, Iran Carlos Lovis. ***Socio-economic reconstruction of the Missões Region/RS.*** Year [?].

Available:<http://www.sober.org.br/palestra/12/10o461.pdf>, Access: Nov,2012.

PEREIRA, Ferdinand C. ***What is Empowerment.*** Sapiência, Informativo Cientifico da FAPEPI,n° 8 Ano III, ISSN 18090915. Jun.2006. Available:< http://www.fapepi.pi.gov.br/novafapepi/sapiencia8/artigos1.php>. Accessed: 20. Mar. 2013.

PORTO. Rogério Ortiz. **Irrigation program, why?** In: Water: source of food. XV Federacite Book. Pg.109-140. Porto Alegre: Federacite, 2007.

PUTNAM, Robert D. ***Comunidade e Democracia:*** *a experiência da Italia moderna.* 3rd Ed. Editora FGV. Rio de Janeiro, 2002.

PROCERGS. ***Management Information - Students, Classes and Schools***. 35th Education Coordination Office, Sao Borja-RS. Available: http://www.webgen.procergs.com.br/cgi-bin/cgigen.exe. Aug,2012.

REA, L. M.; PARKER, R. A. ***Research Methodology:*** *from planning to execution.* Sao Paulo: Pioneira, 2000.

REVISTA INSTITUTO HISTÓRICO E GEOGRAPHICO E ETHOGRAPHICO DO BRASIL. Volume XXVI. Articles by Joao Pedro Gay Canon of Sao Borja. Rio de Janeiro, 1863.

RILLO, Aparicio Silva. **Sao *Borja in questions and answers***: *historical and customs monograph.* Sao Borja: Tricentennial Collection No. 2, 1982.

RIO GRANDE DO SUL. ***School Censuses 2000, 2005 and 2010.*** 35ª Coordenadoria de Educaçao, Sao Borja-RS. Available: Aug,2012.

RODRIGUES, Clàudio Oraindi. ***Sao Borja and its History***. Sao Borja: Tricentennial Collection No. 1, 1982.

ROESCH, Sylvia Maria Azevedo. ***Administration Course Internship Projects: a*** *guide to research, projects, internships and course completion work.* Sao Paulo. Atlas, 2010.

ROSA, Joal de Azambuja. ***References for the Strategic Planning of Agriculture in Rio Grande do Sul***. Article published in the book: Sustainability as a Competitiveness Factor in Agricultural Systems. Pages 15-50. Porto Alegre: Federacite XIX, 2011.

SAINT-HILAIRE, Auguste. ***Journey to Rio Grande do Sul***. Translated by Adroaldo Mesquita da

Costa. 4th Edition. Porto Alegre: Martins Livreiro Editor, 2002.

SEBRAE. **Sebrae publication.** How to carry out market research. Isabela Motta Gomes. Table determining sample size, p.10. : Sebrae, 2005. Available: <02-como_elaborar_P_M.pdf>. Accessed: Nov. 2011.

SEGATTI, S.; HASPANHOL, A. N. **Alternatives for Income Generation on Small Rural Properties.** Article published in the 4th National Meeting for Income Generation in Small Rural Properties. Available:<Altern.GeraçaoRenda.pdf>, Accessed: May, 2012.

SEMPÉ, Moarci Matheus. ***Saint Francis of Borja, the First of the Seven Peoples***. 2nd Ed. Article from the book on Father Francisco Garcia de Prada the Founder. Pages 51 to 66. 1982.

SEN, Amartya. Poverty needs to be tackled at its root. ***Zero Hora***, Economy Supplement, interview with journalists Luiz Antônio Araujo and Maria Isabel Hammes. Page 20, April 23, 2012.

2ND FAMILY FARMING SEMINAR. ***Case: Cooperjeito Caseiro, Sao Francisco de Assis.*** Organizers: Sao Borja City Hall, Emater-Sao Borja. Speaker: Antônio Erico Viero Bem. Sao Borja, July 27, 2012.

SILVA, Cassandra R. O. e.***Metodologia*** e ***Organizaçao do Projeto de Pesquisa:*** guia pràtico. Federal Center for Technological Education of Ceará. Available:< http://www.ufop.br/demet/metodologia.pdf>. Accessed: Mar. 15, 2013. Fortaleza: CEFET, 2004.

SILVEIRA, Hemeterio José Velloso da. ***The Eastern Missions and their Former Dominions***. Porto Alegre: ERUS, 1979.

SOARES, André Luis Ramos; KLAMT, Sérgio Celio. ***Antecedentes Indigenas:*** *Pre- História Compacta do Rio Grande do Sul*. Porto Alegre. Martins Livreiro-Editor, 2005.

SOUTO, A.J.P. ***The town of Sao Francisco de Borja from 1801 to 1834:*** *Luso-Brazilian formation, invasions and ethnic-political-cultural factors.* Seminar 330 years of the History of Sao Borja. Sao Borja: August 17, 2012.

TEIXEIRA, Iberê Athayde. ***1933:*** *The Invasion of Sao Tomé*. Porto Alegre: Corag, 2011.

TIETBOHL, Gilmar. Out of Rhythm: education is still an obstacle in rural areas. **Zero Hora.** Field and Crop Supplement, journalist Nestor Tipa Jùnior. Porto Alegre, October 20, 2012

TORRONTEGUY, Teófilo Otoni Vasconcelos. ***The Origins of Poverty in Rio Grande do Sul.*** State Book Institute. Porto Alegre, 1994.

TUSI DA SILVEIRA et. al. ***Integrated Family Farming Project***. Petrobras Development & Citizenship. URI Santiago/RS. Santiago: 2012.

. ***2nd place in the National Regional Development Award 2010:*** tribute to Celso Furtado. Ministry of National Integration. Technical Report. 2nd Place Title: Integration of fish farming with fertirrigated fruit farming in small rural properties in the Jaguari Valley (RS). URI -Santiago/RS. Pg. 5867.

MINISTÉRIO DA INTEGRAÇAO NACIONALBrasilia, 2010.

VENTURINI, Sérgio. ***Indians in the Missions**: Before, During and After the Jesuits*. Porto Alegre: Martins Livreiro, 2009.

VERGARA, Sylvia Constant. ***People Management.*** 9ª Ed.Sao Paulo: Atlas, 2010.

WAGNER III, John A., HOLLENBECK, John R. ***Organizational Behavior:** creating competitive advantage*. Sao Paulo: Saraiva, 2002.

WAQUIL, Paulo D.;MATTOS, Ely José de. ***Rural and urban poverty in Rio Grande do Sul:** an analysis beyond income.* Article from the project "Public Policies, Family Farming and Rural Poverty in Rio Grande do Sul". FAPERGS, 2002. Available: <Poverty and Income 22_WAQUIL.PDF>, Accessed: July, 2011.

WILKINSON, John. **Markets, networks and values:** the new world of family farming. Porto Alegre: Editora UFRGS, 2008.

ZONTA, Volnei (Org.); Et Al. ***Pedagogy of Alternation and Rural Extension.*** Agricultural Sciences N°1. Frederico Westphalen: URI, 2012.

ZUIN, Luis Fernando Soares; QUEIROZ, Timóteo Ramos. ***Agribusiness:** management and innovation*. Sao Paulo: Saraiva, 2006.

BIBLIOGRAPHY CONSULTED

FURASTÉ, Pedro Augusto. **Normas Técnicas para o Trabalho Cientifico:** Explicitaçao das Normas da ABNT. 16th Ed. Porto Alegre: Dàctilo Plus, 2012.

MARQUES, Pedro Rocha. ***Evaluation of the Competitiveness of Beef Cattle Production Systems in the Western Frontier of Rio Grande do Sul.*** Master's dissertation. Author: Pedro Rocha Marques. Federal University of Rio Grande do Sul, Center for Studies and Research in Agribusiness, Graduate Program in Agribusiness. Porto Alegre: UFRGS, 2010.

ROCHA, Francisco Eduardo de Castro Et Al. ***Application of Content Analysis from Bardin's Perspective in an Evaluation Approach to Pronaf-PB.*** Francisco Eduardo de Castro Rocha ...[et al]. Boletim de pesquisa e desenvolvimento/ Embrapa Cerrados, ISSN 1676-918x;201. Planaltina,DF:Embrapa Cerrados,2008.

APPENDIX

APPENDIX A - Survey form on the socio-productive profiles of small and medium-sized rural properties in the municipality of Sao Borja - RS

1) PROPERTY DATA

Families and workers	Put numbers	
Number of family members	827	
Number of people with a formal contract working on the farm	11	
Number of casual workers per year	85	
Mark with "X"	YES	No
Does the family use a pension to supplement their income?	114 + 36 health	134
Is the family supplying labor to third parties?	57;23,2%	189; 76,8%
Have young people migrated away from the farm?	142	
Is the property the result of family succession?	155,62,5%	90;37,5%
Family Farming	YES	No
Does the property belong to family farming?	211; 85,1%	37:14,9%
The producer has the (DAP) Declaration of Aptitude to Pronaf	84;34,3%	163; 65,7%
Electricity Mark with "X"	YES	No
Your property has electricity	247;99,6%	1;0,4%

Water Dial with "X"	Bad	Regular		Good	Great
How do you rate the quality of drinking water on your property?	13;5,2%	40:16,1%		158; 63,7%	37;14,9%
Mark with "X"	None	Scarce	Regular	Good	Great
How do you rate the amount of water on your property to supply your livestock?	103;41,5%	103;41,5%	35;14,1%	68;27,4%	16;8,5%
Does your property have enough water to irrigate crops?	152;61,3%	38;15,3%	22;8,9%	32;12,9%	4;1,6%
Is your water communal?	YES	168;67,7%	No	80;32,3%	

2) SOCIO-ECONOMIC CLASSIFICATION - Rural Property

Renda Estimada pela Classificação Sócio Econômica Brasil

Categoria		
Categoria nº 1,"E>R$276,70"	17	6,9%
Categoria nº 2,"D>R$484,97"	33	13,3%
Categoria nº 3,"C2>R$726,26"	59	23,8%
Categoria nº 4,"C1>R$1194,53"	58	23,4%
Categoria nº 5,"B2>R$2012,67"	45	18,1%
Categoria nº 6,"B1>R$3479,36"	21	8,5%
Categoria nº 7,"A2>R$6563,73"	7	2,8%
Categoria nº 8,"A1>R$9733,47"	8	3,2%
Total	248	100,0%

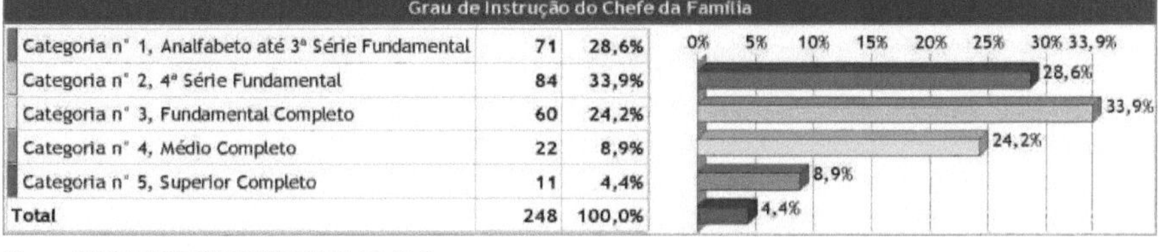

Grau de Instrução do Chefe da Família

Categoria		
Categoria nº 1, Analfabeto até 3ª Série Fundamental	71	28,6%
Categoria nº 2, 4ª Série Fundamental	84	33,9%
Categoria nº 3, Fundamental Completo	60	24,2%
Categoria nº 4, Médio Completo	22	8,9%
Categoria nº 5, Superior Completo	11	4,4%
Total	248	100,0%

3) **LOCATION OF PROPERTY** (district)

4)

	Place in hectares
Total area of properties	5.930,87
Own Area	4.266,07
RENTED AREA	1.361,00
Area in PARTNERSHIP	72,50
Area in CONDOMINIUM	231,30

5) AGRICULTURAL CROPS - (Only fill in what the property grows)

Rice cultivation - TOTAL AREA 150.50 qq	Average yield 168.20 SC/ block
Grain Maize Crop - TOTAL AREA 308.65 ha	Average productivity 22.14 SC/ ha
Maize crop - cob	37 properties; 14.9%
Maize crop - Use in breeding	75 properties; 51.4%
Soya Crop - TOTAL AREA 601 ha	Average Productivity 18.59 SC/ ha
Sorghum Crop - TOTAL AREA 68.5 ha	Average productivity 5.2 SC/ ha
Wheat Crop - TOTAL AREA 437 ha	Average Productivity 25.50 SC/ ha

Other:

6) THE FARM'S AGRICULTURAL CULTURE(S) -

Answer by marking "X"	Handmade	Out of date	Functional	Advanced
The use of technology in farming is	45;18,2%	17;6,9%	58;23,4%	2;0,8%
Answer by marking "X"	Monoculture	Crop rotation	Crop consortium	Planting + animals
Land Management	32;12,9%	30;12,1"%	10;4%	71,28,6%
Answer by marking "X"	Bass	Regular	Good	Great
Use of inputs is	60;24,2%	33;13,3%	20;8,1%	2;0,8%

Answer by marking "X"	Bass	Medium	High
Increasing productivity on your property requires investments that you consider ...	1;0,9 %	31;2 7,2%	82;71, 9%
Your investment condition is	96;84 ,2%	17;1 4,9%	1;0,9 %

Answer by marking "X"	YES	NO
Do you periodically analyze the soil on your property?	31; 12,5%	81; 32,7%
Do you monitor production costs?	21;8,5%	91;36, 7%
Do you monitor sales prices?	21;8,5%	91;36, 7%
Do you monitor the best time to sell?	13;5,2%	99;39, 9%
Do you cut costs without losing quality and productivity?	8;3,2%	103;4 1,5%

Just mark with an "X"	dissatisfied	Slightly dissatisfied	Indifferent	satisfied	Very satisfied
How satisfied are you with the marketing of your produce?	21;8,5%	10;4%	6;2,4%	7;2,8%	1;0,4%

Do you have technical assistance? Yes 28; 11.3% No 100; 40.3%

Who advises? Private company 9, Emater 7, Coptec 5, Agronomist 4, City Hall 3, agricultural technician 1

Frequency of guidance? Monthly 4, on call 4, bimonthly 2, fortnightly 2, quarterly 1, permanent 1

Quality of Guidance? Poor 4, Fair 3, Good 7, Great 2

7) AGRICULTURAL MACHINERY. Do you use it? Mark with "X"

Owned 47;19% Rented 28;11.3% In partnership 29; 11.7% None 60;24.2%

| If you use the agricultural patrol | YES| 69;27.8% | No 94; 37.9%

WITH RESPECT TO OWN AGRICULTURAL MACHINERY, DO YOU CONSIDER: Check the only option with "X".

	Handmade	Out of date	Functional	Advanced
Machinery technology is	3;1,2%	19;7,7%	23;9,3%	2;0,8%
	Low	Regular	Good	Great
Quality in machinery is	6;2,4%	21;8,5%	19; 7,7%	1;0,4%

8) HOW DO YOU COST AGRICULTURAL CROPS: Mark with an "X" and write.

| Resources 120; 48.4% | Other 13; 5.2% | Banks 5; 2% | Own and Banks 11.4.4% |

Own	Companies		
Which bank?	Bank of Brazil 9	Banrisul 5	Sicredi 5

9) SUBSISTENCE CROPS - (Only fill in what the property does). Check "YES" in the box.

Sugarcane cultivation 119;48%	Sweet Potato Crop 175;70.6%	Cassava cultivation 220; 88.7%
Melon cultivation 188; 75%	Watermelon crop 191, 77%	Moranga / Abóboda crop 204; 82.3%
Your main subsistence crop? Write: Cassava Observations: 4 properties have lost all production due to the drought		

Answer by marking "X"	YES	NO
Do you have technical assistance?	19; 7,7%	201; 81,1%

If you answered YES, please answer the questions below.	uestoes:			
Mark the answer with an "X"	Technical Agriculture	Agronomist	Emater	Coptec
Who provides assistance?	2;0,8%	1; 0,4%	7; 2,8%	9;3,6%

	Weekly	Monthly	Quarterly	Half-yearly
The frequency of assistance is	1,0,4%	8;3,2%	5,2%	5,2%
	Low	Regular	Good	Great
Quality of care è	2; 0,8%	9; 3,6%	6; 2,4%	1; 0,4%

Subsistence crop marketing the main one

Answer by marking "X"	YES	NO
Do you sell your produce?	60; 24,2%	160; 64,5%

If Yes, Answer	Direct to Consumer	Regional Wholesaler	Regional Supermarket	Other
Where do you sell?	53; 21,4%	1; 0,4%	1;0,4%	10; 4%

Other? Fairs, mini-markets, steakhouses, etc.	school meals	
Answer by marking "X"	YES	NO
Is it easy to market?	44; 17,7%	15;6,1%

Just mark with an "X"	dissatisfied	Slightly dissatisfied	Indifferent	satisfied	Very satisfied
How satisfied are you with the marketing of your produce?	1;0,4%	15;6,%	4;1,6%	35;14,1%	4;1,6%

10) VEGETABLES; Mark with an "X" if yes and describe the vegetables.

Yes Produces 206; 83,1% Which ones? Lettuce 195; 16.8% cabbage 142; 12.2%, carrots 130; 11.2%, beet 114; 9.8%			
Answer by marking "X"	Low	Average	High
The use of technology to produce vegetables on the farm is?	162;65,3%	47;19%	2;0,8%

Do you compost?	YES	57;23%	NO	154;62,1%
Do you use chemical fertilizers?	YES	13,5,2%	NO	198;79,8%
Do you use organic fertilizer?	YES	210;84,7%	NO	1;0,4%
Do you sell your produce?	YES	53;21,4%	NO	158;63,7%

If Yes, Answer	Direct to Consumer	Other		
Where do you sell?	50;20,2%	20;8%		

Other? Fairs, mini-market, public market		
Answer by marking "X"	YES	NO
Is it easy to sell?	40;16,1%	11;4,4%

Just mark with an "X"	dissatisfied	Slightly dissatisfied	Indifferent	satisfied	Very satisfied
How satisfied are you with the marketing of your produce?	2;0,8%	17;6,9%	2;0,8%	31;12,5%	1;0,4%

11) PASTURE: Fill in the corresponding areas in hectares, if any. And describe the ones that are carried out.

Native Grassland	2.622,25	ha	
Winter grazing	622,25	ha	Which ones?ryegrass 127; 51.4%, black oats 116; 47%
Summer grazing	375,50	ha	Which ones?summer oats 39; 25.8%, millet 30; 19.9%, sorghum 22;14.6%
Perennial pasture	423,60	ha	Which ones? brachiaria 78; 49.4%, tifton 34;21.5%, sugar cane 8; 5.1%

12) Large Animals: Fill in all the numbers (animals); mark with an "X" the types of own consumption and sales that are carried out per number of animals per year.

Cattle: TOTAL Animals 3,262 head. Own consumption (yes) 103; 41.5%	Annual cattle marketing: Number of animals Adults 386 Young people 411
Bubalinos: TOTAL Animals 1 head (Búfalos) Own consumption (yes)	Annual marketing of buffaloes: Number of animals
Equines: TOTAL Animals 172 cab.	Marketing of horses: Number of animals Adults 6 cab.

Where do you sell?

	Direct to the Producer	Via Intermediaries	By Marchante	Home of Meat/Mini-markets	Frigorificos da Regiao	Other Refrigerators Regions
Cattle	50;20,2%	28;11,3%	6;2,4%	5;2%	2;0,8%	0;0,0%
Bubalinos				1;0,4%		

Just mark with an "X"	dissatisfied	Slightly dissatisfied	Indifferent	satisfied	Very satisfied
How satisfied are you with the marketing of your produce?	0;0,0%	28;11,3%	8;3,2%	35;14,1%	2;0,8%

How do you pay for large animals? Mark an "X".

Own resources 97;39.1%	Banks 1;0.4%	Own and Banks 7;2.8%
Which bank? Bank of Brazil 6	Banrisul 1	Sicredi 1

How animal health (health control) is carried out.

Answer YES or NO	YES	NO
Is there a pre-established health calendar?	15;6,1%	87;35,1%
Are vaccinations given for the main endemic diseases in the region?	102;41,1%	0;0,0%
Do the animals receive regular treatment for endo- and ecto-parasites?	101;40,7%	1;0,4%
Is there differentiated health management by animal category (age)?	27;10,9%	75;30,2%

MILK SIZE Fill in numbers, mark "X" and numbers

Number of Animals 1,565 cab.	Total Daily Milk Production 5.011 liters
Predominant breeds:	
Dutch	Number of animals 705
Jersey	Number of animals 156
Swiss Brown	Number of animals 1
Crosses	Number of animals 636

Other? 67

How animal health (sanitary control) is carried out at TAMBO DE LEITE.

Answer by marking "X"	YES	NO
Is there a pre-established health calendar?	28;11,3%	73,29,4%
Are vaccinations given for the main endemic diseases in the region?	99;39,9%	3;1,2%
Do the animals receive regular treatment for endo- and ecto-parasites?	99;39,9%	3;1,2%
Is there differentiated health management by animal category (age)?	44;17,7%	58;23,4%

Technical assistance for the Milk Drum

Answer by marking "X"	YES	NO
Does a professional (Med.Vet, Zoot, Agron) visit the property periodically?	13;5,2%	87;35,1%
Emater advises YES 15; 6.1% or Coptec, advises YES 26;10.5%	-	-
Are there any one-off services? (ringing, andrology, grazing...)	29;11,7%	71;28,6%
Do you prioritize technical assistance when buying inputs?	11;4,4%	89;35,9%

Do you supplement TAMBO DE LEITE animals?

Answer by marking "X" | YES NO|

Does it supplement milk?	81;32,7%	19;7,7%
If YES, keep answering	YES	NO
Do you supplement with crop residues?	74;29,8%	9; 3,6%
Do you supplement with industrial feed?	50;20,2%	32;12,9%
Do you use hay as a food supplement?	7;2,8%	75;30,2%
Do you use silage as a feed supplement?	16;6,5%	66;26,6%

Method of marketing in the MILK TAMBLE. Mark with an "X".

Sale of a private company (integration) 37; 14.9%

Sale to cooperative 4; 1.6% *OBS: Cooperatives outside the municipality

Direct sales to consumers 44; 17.7%

Receipt of marketing in the MILK TAMBLE. Mark with an "X".

Daily 27;10,9% Weekly 15;6.1% Fortnightly 16;6.5% Monthly 30;12,1%

Just mark with an "X"	dissatisfied	Slightly dissatisfied	Indifferent	satisfied	A lot satisfied
How satisfied are you with the marketing of your produce?	8;3,2%	17;6,9%	9;3,6 %	30;1 2,1%	3;1,2%

Do you use machinery in the MILK TAMBLE?

Owned 31;12.5% Rented 0 In partnership 3;1.2% None 50;20.2%

If the answer is Own, still Mark the options with an "X" below.

Mark an answer	Handmade	Out of date	Functional	Advanced
Machinery technology is	1;0,4%	2;0,8%	18;7,3%	11;4,4%
	Low	Regular	Good	Great

Quality in machinery is	2;0,8%	2;0,8%	19;7,7%	9;3,6%

How the milk drum is paid for

Own resources 81;32.7%	Banks 1;0.4%	Own and Banks 11;4.4%
Which bank?	Bank of Brazil 8	Sicredi 2

OBS: 7 properties have given up tambo farming.

13) Small Animals: Fill in the numbers there are (animals); mark with "X" the types of own consumption and sales carried out number of animals per year

Sheep: TOTAL Animals 1,793 head. Own consumption (yes) 74;29.8%	Annual sheep marketing: Number of animals Adults 144 Youth 184
Pigs: TOTAL Animals 708 cab. Own consumption (yes) 121; 48.8%	Annual pig marketing: Number of animals Adults 39 cab. Youth 40

Which SMALL animal generates the MOST income on the property?

Write: Pigs 27; 52.9%

(continued) Where do you sell SMALL ANIMALS?

	Direct to Producer/	Via Intermediaries	Marchante	Home of Meat/Mini-markets	Frigorificos da Regiao	Consumer
Sheep	18;7,3%	4;1,6%	0;0,0%	1;0,4%	0;0,0%	15;6,1%
Pigs	14;5,7%	1;0,4%	0;0,0%	0;0,0%	0;0,0%	20;8,1%

Others?					
Just mark with an "X"	dissatisfied	Slightly dissatisfied	Indifferent	satisfied	Very satisfied
How satisfied are you with the marketing of your produce?	0;0,0%	8;3,2%	7;2,8%	30;12,1%	2;0,8%

How do you pay for SMALL animals? Mark an "X".

Own resources 96; 49.5%	Banks 1;0.4%	Own and Banks
Which bank? Bank of Brazil 1		

How animal health (health control) is carried out on SMALL animals.

Answer YES or NO	YES	NO
Is there a pre-established health calendar?	7;2,8%	71;28,6%
Are vaccinations carried out for the main endemic diseases in the region?	56;22,6%	24;9,7%
Is there differentiated health management by animal category (age)?	53;21,4%	27;10,9%

Technical assistance for SMALL animals

Answer by marking "X"	YES	NO
Does a professional (Med.Vet, Zoot, Agron) visit the property periodically?	9;3,6%	71;28,6%
Does Emater provide advice?	7;2,8%	50;20,2%
Does Coptec provide advice?	7;2,8%	15;6,1%
Are there any one-off services (ringing, andrology, grazing)?	15;6,1%	63;25,4%
Do you prioritize technical assistance when buying inputs?	5;2%	74;29,8%

14) Birds. Mark with an "X" the options that occur.

Poultry on the property?	YES 213;85.2%	NO 3,1,2%
Which ones?	Own Consumption	Marketing
Hens and roosters 213; 85.2%	208;83,9%	Animals 40;16.1% Eggs 77;31.1%
Angolan chickens 13;5.2%		

Patos 16;6.5%		
Perus 4;1.6%		

Just mark with an "X"	dissatisfied	Slightly dissatisfied	Indifferent	satisfied	Very satisfied
How satisfied are you with the marketing of your produce?	4;1,6%	16;6,5%	17; 6,9%	39;1 5,7%	4;1,6%

15) Aquaculture - Fish farming and marketing

How many dams and ponds do you own? 23

Which fish do you breed?

Describe: Carp 20;33.3%, jundià 6;10%, pacu 4; 6.7%

Commercialize? 3 properties

Just mark with an "X"	dissatisfied	Slightly dissatisfied	Indifferent	satisfied	Very satisfied
How satisfied are you with the marketing of your produce?	0;0,0%	1,0,4%	1,0,4%	1,0,4%	0;0,0%

16) (continued) Beekeeping - Bees

How many boxes do you have? 505 boxes

What is your honey production per year? 5,371 kg

How do you market honey and other products? Direct consumer 23; 23%, through partnerships 10; 10%

Just mark with an "X"	dissatisfied	Slightly dissatisfied	Indifferent	satisfied	Very satisfied
How satisfied are you with the marketing of your produce?	0;0,0%	1;0,4%	3;1,2%	25;10,1%	0;0,0%

17) Forestry - Afforestation

Which species?

Native forest 122; 29.7%, eucalyptus 120; 29.2%, cinnamon 21; 5.1%

18) Orchard - Fruit growing

Which fruit species? Orange 201;21.8%, bergamot 190; 20.6%, peach 113;12.3%, mango 46;5%

What area is used for the orchard? 40.52 ha

How do you sell? Final consumer 25; 11.9%, fair 5; 2.4%

Who provides guidance? Emater 11;6.9%, Coptec 10;6.3%, city hall 3; 1.9%

Just mark with an "X"	dissatisfied	Slightly dissatisfied	Indifferent	satisfied	Very satisfied
How satisfied are you with the marketing of your produce?	0;0,0%	6;2,4%	2;0,8%	30;12,1%	3;1,2%

19) Are you a member of any cooperative or producers' union?

Yes 86;34.7%	No 152;61.3%	Which? Cotrisal 42; 21.9%, associations 28; 14.6%, settlements 8;7.3%

What's your opinion?

Mark with an "X"	Strongly disagree	I disagree	Indifferent	I agree	I agree Absolutely
Collective marketing	7;2,8%	11;4,4%	135;54,4%	76;30,7%	8;3,2%
Group buying	9;3,6%	8;3,2%	151;60,9%	64;25,8%	7;2,8%
Sales to cooperatives	5;2%	15;6,1%	146;58,9%	66;26,6%	7;2,8%
Sales to Associations	3;1,2%	4;1,6%	171;69%	57;23%	4;1,6%

20) List in descending order of importance what the property NEEDS for a greater rural competitiveness. Quotes:

Irrigation 97 properties, technical assistance 83 properties, financial resources 79 properties

Sugar 73 properties, soil correction 59 properties, artesian well 35 properties, marketing 34 properties

21) List in descending order of importance what THREATENS the property in rural competitiveness

Quotes:.

Drought, 181 properties, lack of income 120 properties, aerial herbicide 97 properties, lack of marketing 48 properties, cattle rustling 32 properties, indebtedness 26 properties,

22) What short courses would be of interest to your family or rural property?

a maximum of three options). Quotes:

rural administration 37;14.9%	Bovine farming 59;23,8%	people management 4;1,6%	legislation legislation agro-industry 4; 1.6%
organic farming 23;9,35	rural entrepreneurship 28;11.3%	food processing and canning 34;13.7%	rural marketing 3;1,2%
Beekeeping 12;4.8%	Fruit growing 25;10.1%	basic computer science 14;5,7%	sheep farming 17;6,9%
food waste utilization 32;12,9%	rural cash flow 3;1.2%	Irrigation 60;24.2%	Pig farming 9;3.6%
Poultry farming 25;10,1%	environmental management 2;0,8%	Horticulture 29;11.7%	Fish farming 15; 6.1%

Other: 8; 3.2% dairy farming and bovine insemination

23) Homemade Products

Answer by marking "X"	YES	NO
Do you sell homemade products?	77;31,1%	163;65,7%
Which properties? Milk 32 properties, eggs 25 properties, cheese 24 properties, sweets 24 properties,		

Cuca 17 properties, rapadura 14 properties

ANNEX

ANNEX 1 - Survey form for the analysis of the influence of human and social capacity on local development - Sao Borja, RS.

Quantitative interviews N°Municipality:. Date: _/_/

Interviewee:. Education: Location:

In general, would you say that you can trust most of the people and institutions in your community?

Confidence	All ( )**10** ( ) **9** ( ) **8** ( ) **7** ( ) **6** ( ) **5** ( ) **4** ( ) **3** ( ) **2** ( ) **1** None ( ) **0**

In a community, people lead each other to bring about development. On a scale of (zero) to (ten), what is the attitude of the leaders in your community towards development?

Leadership	Maximum quantity ( )**10** ( ) **9** ( ) **8** ( ) **7** ( ) **6** ( ) **5** ( ) **4** ( ) **3** ( ) **2** ( ) **1** Absence ( ) **0**

In the community, some people are entrepreneurs, while others are not. In general, in relation to your community, would you say that there is entrepreneurship?

Entrepreneurship	Maximum quantity ( )**10** ( ) **9** ( ) **8** ( ) **7** ( ) **6** ( ) **5** ( ) **4** ( ) **3** ( ) **2** ( ) **1** Absence ( ) **0**

ANNEX 2 - Recommendation from the Central Committee of the Festivities for the 330th Anniversary of the Establishment of the Municipality of Sao Borja for the "Origona or Origone Cheese" project.

COMISSÃO DOS FESTEJOS DOS 330 ANOS DE
INSTALAÇÃO DO MUNICÍPIO DE SÃO BORJA

330
São Borja

DECLARAÇÃO

A Comissão Central dos Festejos dos 330 Anos de Instalação do Município de São Borja vem por meio desta recomendar o projeto do "Queijo Origona ou Origone" a ser desenvolvido a posteriori.

Projeto este apresentado neste momento pelo Pesquisador Professor Augusto José Fisto Souto que consta dos assuntos da sua dissertação de Mestrado.

São Borja, 28 de Novembro de 2012.

Rodrigo Ba[illegible]
Presidente

FSC
www.fsc.org
MIX
Papier aus verantwortungsvollen Quellen
Paper from responsible sources
FSC® C105338

Printed by Books on Demand GmbH, Norderstedt / Germany